Viv Tolley

ESSENTIALS

AQA

GCSE Science B

Revision Guide

Contents

Revised

4 How Science Works Overview

11 Data Sheets

Unit 1: My World

My Wider World

12 Our Changing Universe (3.3.1.1)

14 Our Changing Planet (3.3.1.2)

20 Materials Our Planet Provides (3.3.1.3)

29 Using Materials from Our Planet to Make Products (3.3.1.4)

Life on Our Planet

34 Life on Our Planet (3.3.2.1)

37 Biomass and Energy Flow Through the Biosphere (3.3.2.2)

39 The Importance of Carbon (3.3.2.3)

40 Exam Practice Questions

Unit 2: My Family and Home

My Family

42 Control of Body Systems (3.4.1.1)

46 Chemistry in Action in the Body (3.4.1.2)

48 Human Inheritance and Genetic Disorders (3.4.1.3)

My Home

53 Materials Used to Construct Our Homes (3.4.2.1)

58 Fuels for Cooking, Heating and Transport (3.4.2.2)

59 Generation and Distribution of Electricity (3.4.2.3)

My Property

64 The Cost of Running Appliances in the Home (3.4.3.1)

66 Electromagnetic Waves in the Home (3.4.3.2)

68 Exam Practice Questions

Contents

Revised

Unit 3: Making My World a Better Place

Improving Health and Wellbeing

70 The Use (and Misuse) of Drugs (3.5.1.1)

73 The Use of Vaccines (3.5.1.2)

74 The Use of Ionising Radiation in Medicine (3.5.1.3)

Making and Improving Products

77 Uses of Electroplating (3.5.2.1)

79 Developing New Products (3.5.2.2)

81 Selective Breeding and Genetic Engineering (3.5.2.3)

Improving our Environment

83 Environmental Concerns – Making and Using Products (3.5.3.1)

87 Saving Energy in the Home (3.5.3.2)

89 Controlling Pollution in the Home (3.5.3.3)

90 Exam Practice Questions

92 Answers

94 Glossary

95 Index

96 Electronic Structure of the First 20 Elements

IBC Periodic Table

N.B. The numbers in brackets correspond to the reference numbers on the AQA GCSE Science B specification.

How Science Works Overview

How Science Works – Explanation

The AQA GCSE Science B specification incorporates:

- **Science Content** – all the scientific explanations and evidence that you need to know for the exams. (It is covered on pages 12–89 of this revision guide.)
- **How Science Works** – a set of key concepts, relevant to all areas of science. It covers…
  - the relationship between scientific evidence, and scientific explanations and theories
  - how scientific evidence is collected
  - how reliable and valid scientific evidence is
  - the role of science in society
  - the impact science has on our lives
  - how decisions are made about the ways science and technology are used in different situations, and the factors affecting these decisions.

Your teacher(s) will have taught these two types of content together in your science lessons. Likewise, the questions on your exam papers will probably combine elements from both types of content. So, to answer them, you'll need to recall and apply the relevant scientific facts and knowledge of how science works.

The key concepts of How Science Works are summarised in this section of the revision guide (pages 4–10). You should be familiar with all of these concepts. If there is anything you are unsure about, ask your teacher to explain it to you.

How Science Works is designed to help you learn about and understand the practical side of science. It aims to help you develop your skills when it comes to…

- evaluating information
- developing arguments
- drawing conclusions.

The Thinking Behind Science

Science attempts to explain the world we live in.

Scientists carry out investigations and collect evidence in order to…

- **explain phenomena** (i.e. how and why things happen)
- **solve problems** using evidence.

Scientific knowledge and understanding can lead to the **development of new technologies** (e.g. in medicine and industry), which have a huge impact on **society** and the **environment**.

The Purpose of Evidence

Scientific evidence provides **facts** that help to answer a specific question and either **support** or **disprove** an idea or theory. Evidence is often based on data that has been collected through **observations** and **measurements**.

To allow scientists to reach conclusions, evidence must be…

- **repeatable** – other people should be able to repeat the same process
- **reproducible** – other people should be able to reproduce the same results
- **valid** – it must be repeatable, reproducible and answer the question.

N.B. If data isn't repeatable and reproducible, it can't be valid.

To ensure scientific evidence is repeatable, reproducible and valid, scientists look at ideas relating to…

- observations
- investigations
- measurements
- data presentation
- conclusions and evaluation.

Observations

Most scientific investigations begin with an **observation**. A scientist observes an event or phenomenon and decides to find out more about how and why it happens.

The first step is to develop a **hypothesis**, which suggests an explanation for the phenomenon. Hypotheses normally suggest a relationship between two or more **variables** (factors that change).

Hypotheses are based on…
- careful observations
- existing scientific knowledge
- some creative thinking.

The hypothesis is used to make a **prediction**, which can be tested through scientific investigation. The data collected from the investigation will…
- support the hypothesis **or**
- show it to be untrue (refute it) **or**
- lead to the modification of the original hypothesis or the development of a new hypothesis.

If the hypothesis and models we have available to us do not completely match our data or observations, we need to check the validity of our observations or data, or amend the models.

Sometimes, if the new observations and data are valid, existing theories and explanations have to be revised or amended, and so scientific knowledge grows and develops.

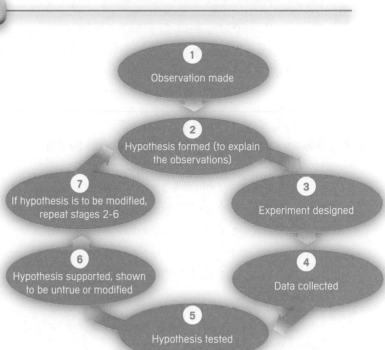

1. Observation made
2. Hypothesis formed (to explain the observations)
3. Experiment designed
4. Data collected
5. Hypothesis tested
6. Hypothesis supported, shown to be untrue or modified
7. If hypothesis is to be modified, repeat stages 2-6

Example

- Two scientists **observe** that freshwater shrimp are only found in certain parts of a stream.
- They use scientific knowledge of shrimp and water flow to develop a **hypothesis**, which relates the presence of shrimp (dependent variable) to the rate of water flow (independent variable). For example, a hypothesis could be: the faster the water flows, the fewer shrimp are found.
- They **predict** that shrimp are only found in parts of the stream where the water flow rate is below a certain value.
- They **investigate** by counting and recording the number of shrimp in different parts of the stream, where water flow rates differ.
- The **data** shows that more shrimp are present in parts of the stream where the flow rate is below a certain value. So, the data **supports** the hypothesis. But, it also shows that shrimp aren't always present in these parts of the stream.
- The scientists realise there must be another factor affecting the distribution of shrimp. They **refine their hypothesis**.

How Science Works Overview

An **investigation** involves collecting data to find out whether there is a relationship between two **variables**. A variable is a factor that can take different values.

In an investigation there are two types of variables:
- **Independent** variable – can be changed by the person carrying out the investigation. For example, the amount of water a plant receives.
- **Dependent** variable – measured each time a change is made to the independent variable, to see if it also changes. For example, the growth of the plant (measured by recording the number of leaves).

For a measurement to be valid it must measure only the appropriate variable.

Variables can have different types of values:
- **Continuous variables** – can take any numerical value (including decimals). These are usually measurements, e.g. temperature.
- **Categoric variables** – a value described by a label, usually a word, e.g. different breeds of dog or blood group.
  - **Discrete variables** – only take whole-number values. These are usually quantities, e.g. the number of shrimp in a stream.
  - **Ordered variables** – have relative values, e.g. 'small', 'medium' or 'large'.

N.B. Numerical values, such as continuous variables, tend to be more informative than ordered and categoric variables.

An investigation tries to find out whether an **observed** link between two variables is…
- **causal** – a change in one variable causes a change in the other, e.g. the more cigarettes you smoke, the greater the chance that you will develop lung cancer.
- **due to association** – the changes in the two variables are linked by a third variable, e.g. as grassland decreases, the number of predators decreases (caused by a third variable, i.e. the number of prey decreasing).
- **due to chance** – the change in the two variables is unrelated; it is coincidental, e.g. people who eat more cheese than others watch more television.

Controlling Variables

In a **fair test**, the only factor that should affect the dependent variable is the independent variable. Other **outside variables** that could influence the results are kept the same (control variables) or eliminated.

It's a lot easier to control all the other variables in a laboratory than in the field, where conditions can't always be controlled. The impact of an outside variable (e.g. light intensity or rainfall) has to be reduced by ensuring all the measurements are affected by it in the same way. For example, all the measurements should be taken at the same time of day.

Control groups are often used in biological and medical research to make sure that any observed results are due to changes in the independent variable only.

A sample is chosen that 'matches' the test group as closely as possible except for the variable that is being investigated, e.g. testing the effect of a drug on reducing blood pressure. The control group must be the same age, gender, have similar diets, lifestyles, blood pressure, general health, etc.

Investigations (Cont.)

Accuracy and Precision

How accurate data needs to be depends on what the investigation is trying to find out. For example, when measuring the volume of acid needed to neutralise an alkaline solution it is important that equipment is used that is able to accurately and precisely measure volumes of liquids.

The data collected must be **precise** enough to form a **valid conclusion**: it should provide clear evidence for or against the hypothesis.

Measurements

Apart from control variables, there are a number of factors that can affect the reliability and validity of measurements:

- **Accuracy of instruments** – depends on how accurately the instrument has been calibrated. An accurate measurement is one that is close to the true value.
- **Resolution (or sensitivity) of instruments** – determined by the smallest change in value that the instrument can detect. The more sensitive the instrument, the more **precise** the value. For example, bathroom scales aren't sensitive enough to detect changes in a baby's mass, but the scales used by a midwife are.
- **Human error** – even if an instrument is used correctly, human error can produce random differences in repeated readings or a systematic shift from the true value if you lose concentration or make the same mistake repeatedly.
- **Systematic error** – can result from repeatedly carrying out the process incorrectly, making the same mistake each time.
- **Random error** – can result from carrying out a process incorrectly on odd occasions or by fluctuations in a reading. The smaller the random error the greater the accuracy of the reading.

To ensure data is as accurate as possible, you can…
- calculate the **mean** (average) of a set of repeated measurements to reduce the effect of random errors
- increase the number of measurements taken to improve the reliability of the mean / spot anomalies.

Preliminary Investigations

A trial run of an investigation will help identify appropriate values to be recorded, such as the number of repeated readings needed and their range and interval.

You need to examine any **anomalous** (irregular) values to try to determine why they appear. If they have been caused by equipment failure or human error, it is common practice to ignore them and not use them in any calculations.

There will always be some variation in the actual value of a variable, no matter how hard we try to repeat an event.

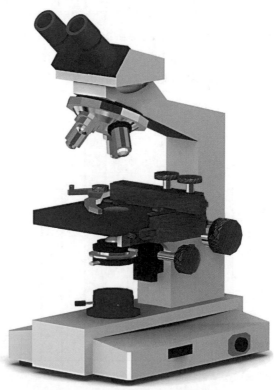

How Science Works Overview

Presenting Data

Data is often presented in a **chart** or **graph** because it makes…

- any patterns more obvious
- it easier to see the relationship between two variables.

The **mean** (or average) of data is calculated by adding all the measurements together, then dividing by the number of measurements taken:

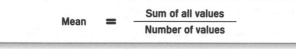

$$\text{Mean} = \frac{\text{Sum of all values}}{\text{Number of values}}$$

If you present data clearly, it is easier to identify any anomalous (irregular) values. The type of chart or graph you use to present data depends on the type of variable involved:

1 **Tables** organise data (but patterns and anomalies aren't always obvious)

Height of student (cm)	127	165	149	147	155	161	154	138	145
Shoe size	5	8	5	6	5	5	6	4	5

2 **Bar charts** display data when the independent variable is categoric or discrete and the dependent variable is continuous.

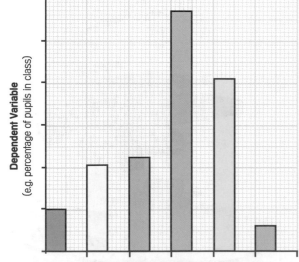

Independent Variable (e.g. eye colour)

3 **Line graphs** display data when both variables are continuous.

- Points are joined by straight lines if you don't have data to support the values between the points.
- A line of best fit is drawn if there is sufficient data or if a trend can be assumed.

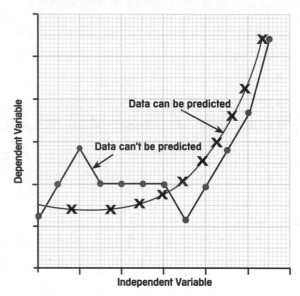

4 **Scattergrams** (scatter diagrams) show the underlying relationship between two variables. This can be made clearer if you include a **line of best fit**. A line of best fit could be a straight line or a smooth curve.

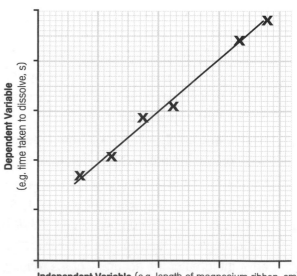

Independent Variable (e.g. length of magnesium ribbon, cm)

Conclusions and Evaluations

Conclusions **should**…
- describe patterns and relationships between variables
- take all the data into account
- make direct reference to the original hypothesis or prediction
- try to explain the results / observations by making reference to the hypothesis as appropriate.

Conclusions **should not**…
- be influenced by anything other than the data collected (i.e. be biased)
- disregard any data (except anomalous values)
- include any unreasoned speculation.

An **evaluation** looks at the whole investigation. It should consider…
- the original purpose of the investigation
- the appropriateness of the methods and techniques used
- the reliability and validity of the data
- the validity of the conclusions.

The **reliability** of an investigation can be increased by…
- looking at relevant data from secondary sources (i.e. sources created by someone who did not experience first hand or participate in the original experiment)
- using an alternative method to check results
- ensuring results can be reproduced by others.

Science and Society

Scientific understanding can lead to technological developments. These developments can be exploited by different groups of people for different reasons. For example, the successful development of a new drug…
- benefits the drugs company financially
- improves the quality of life for patients
- can benefit society (e.g. if a new drug works, then maybe fewer people will be in hospital, which reduces time off sick, cost to the NHS, etc).

Scientific developments can raise certain **issues**. An issue is an important question that is in dispute and needs to be settled. The resolution of an issue may not be based on scientific evidence alone.

There are several different types of **issue** that can arise:
- **Social** – the impact on the human population of a community, city, country, or the world.
- **Economic** – money and related factors like employment and the distribution of resources.
- **Environmental** – the impact on the planet, its natural ecosystems and resources.
- **Ethical** – what is morally right or wrong; requires a value judgement to be made.

N.B. There is often an overlap between social and economic issues.

Peer Review

Peer review is a process of self-regulation involving qualified professional individuals or experts in a particular field who examine the work undertaken critically. The vast majority of peer review methods are designed to maintain standards and provide credibility for the work that has been undertaken. These methods vary depending on the nature of the work and also on the overall purpose behind the review process.

How Science Works Overview

Evaluating Information

It is important to be able to evaluate information relating to social-scientific issues, for both your GCSE course and to help you make informed decisions in life.

When evaluating information…
- make a list of **pluses** (pros)
- make a list of **minuses** (cons)
- consider how each point might **impact on society**.

You also need to consider whether the source of information is reliable and credible. Some important factors to consider are…
- **opinions**
- **bias**
- **weight of evidence**.

Opinions are personal viewpoints. Opinions backed up by valid and reliable evidence carry far more weight than those based on non-scientific ideas.

Opinions of experts can also carry more weight than non-experts.

Information is **biased** if it favours one particular viewpoint without providing a balanced account.

Biased information might include incomplete evidence or try to influence how you interpret the evidence.

Scientific evidence can be given **undue weight** or dismissed too quickly due to…
- political significance (consequences of the evidence could provoke public or political unrest)
- status of the experiment (e.g. if they do not have academic or professional status, experience, authority or reputation).

Limitations of Science

Although science can help us in lots of ways, it can't supply all the answers. We are still finding out about things and developing our scientific knowledge.

There are some questions that science can't answer. These tend to be questions…
- where beliefs, opinions and ethics are important
- where we don't have enough reproducible, repeatable or valid evidence.

Science can often tell us if something **can** be done, and **how** it can be done, but it can't tell us whether it **should** be done.

Decisions are made by individuals and by society on issues relating to science and technology.

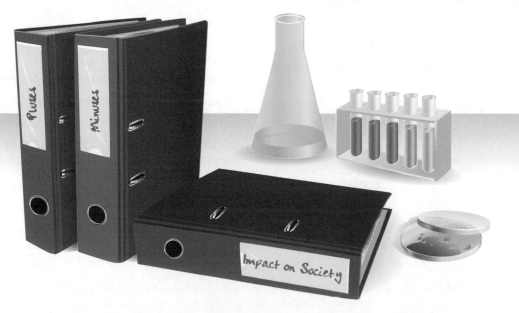

Quantities, Units and Symbols

You will need to know how to measure and / or calculate the following quantities, using the correct units and their symbols:

Quantity	Units / symbols
mass	kilogram, kg; gram, g; milligram, mg; microgram, μg
length	kilometre, km; metre, m; centimetre, cm; millimetre, mm; micrometre, μm
volume	cubic metre, m^3; cubic decimetre, dm^3 (litre, l); cubic centimetre, cm^3 (millilitre, ml)
time	hour, h; minute, min; second, s
temperature	degrees Celsius, $^{\circ}$C
chemical quantity	mole, mol
potential difference (voltage)	volt, V
current	ampere, A; milliampere, mA
force	newton, N
energy / work	kilojoule, kJ; joule, J; kilowatt-hour, kWh
power	kilowatt, kW; watt, W
frequency	hertz, Hz
wavelength	metre, m
velocity (wave speed)	metre per second, m/s

Chemical Symbols – Elements

You will need to know the chemical symbols for the first 20 elements in the Periodic Table. You also need to know the chemical symbols of the following elements.

Metal	Chemical Symbol
Copper	Cu
Gold	Au
Iron	Fe
Lead	Pb
Nickel	Ni
Silver	Ag
Zinc	Zn

Name of Element	Atomic Number	Symbol
Hydrogen	1	H
Helium	2	He
Lithium	3	Li
Beryllium	4	Be
Boron	5	B
Carbon	6	C
Nitrogen	7	N
Oxygen	8	O
Fluorine	9	F
Neon	10	Ne
Sodium	11	Na
Magnesium	12	Mg
Aluminium	13	Al
Silicon	14	Si
Phosphorus	15	P
Sulfur	16	S
Chlorine	17	Cl
Argon	18	Ar
Potassium	19	K
Calcium	20	Ca

Chemical Formulae – Compounds

You will need to know these names and formulae.

Compound	Formula
Aluminium hydroxide	$Al(OH)_3$
Ammonia	NH_3
Calcium carbonate	$CaCO_3$
Calcium hydroxide	$Ca(OH)_2$
Calcium oxide	CaO
Carbon dioxide	CO_2
Chlorine gas	Cl_2
Hydrogen gas	H_2
Hydrogen chloride gas	HCl
Magnesium hydroxide	$Mg(OH)_2$
Methane	CH_4
Nitrogen gas	N_2
Oxygen gas	O_2
Sodium hydrogencarbonate	$NaHCO_3$
Water	H_2O

U1 Our Changing Universe

Observing the Universe

Observations of the **solar system** and the **galaxies** in the Universe can be carried out…
- on Earth
- from space.

One method of observation is to use a **telescope**.

Different types of telescope can detect visible light or other electromagnetic radiations, e.g. radio waves or X-rays.

These observations provide evidence for changes that are taking place in the Universe.

Telescopes

A **reflecting telescope reflects** light using **mirrors**. It can only be used at night.

A **refracting telescope refracts** light at each end using **lenses**. It produces sharp, detailed images but can only be used at night.

Radio telescopes pick up **radio waves** instead of light waves. The radio waves are emitted by objects in space.

Radio waves have a larger wavelength than light waves. So in order to receive good signals, radio telescopes need large **antennae** or lots of smaller antennae working together.

Most radio telescopes use a **parabolic dish** to reflect the radio waves to a **receiver**, which detects and amplifies the signal.

X-ray telescopes pick up X-rays from space.

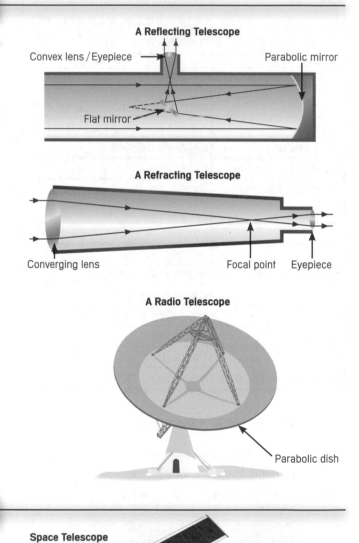

A Reflecting Telescope

Convex lens / Eyepiece   Parabolic mirror

Flat mirror

A Refracting Telescope

Converging lens   Focal point   Eyepiece

A Radio Telescope

Parabolic dish

Using Telescopes

Observing the Universe **from the Earth** is limited by the **atmosphere**. Interference from clouds, weather storms or light pollution reduces the quality of images, so many telescopes are placed on the top of mountains and in areas with low levels of pollution.

Putting telescopes **into space** means that they orbit the Earth outside the atmosphere. The images produced aren't affected by atmospheric interference.

Space Telescope

Key Words   Telescope • Reflection • Refraction • X-ray

Red-shift

If a wave source is moving away from or towards an observer, there will be a change in the...

- observed **wavelength**
- observed **frequency**.

The model that is used to describe this phenomenon is known as the **Doppler effect**. An ambulance racing past you is a good example of the Doppler effect with sound waves:

- When the source moves **away** from you, the observed **wavelength increases** and the frequency decreases. This is known as **red-shift**. The wavelengths 'shift' towards the red end of the **electromagnetic spectrum**.
- When the source moves **towards** you, the observed **wavelength decreases** and the **frequency increases**.

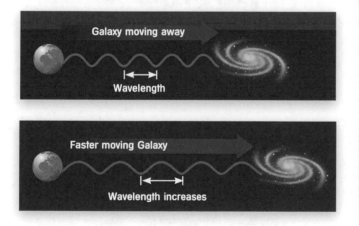

The light observed from most galaxies in the Universe is **red-shifted**. In fact the further away a **galaxy** is, the faster it is moving and the bigger the observed increase in wavelength.

This observed red-shift suggests that...

- the whole Universe is **expanding**
- the Universe might have started billions of years ago, from one small place, with a huge explosion known as the '**Big Bang**'.

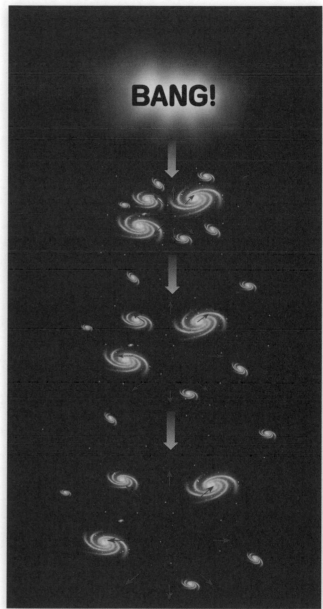

BANG!

Quick Test

1. Name four types of telescope that can be used to observe the Universe.
2. How does a refracting telescope refract light?
3. Give one advantage of observing the Universe from telescopes in space.
4. When a wave source moves away from us, what happens to the wavelength of the light in its spectrum?

U1 Our Changing Planet

Structure of the Earth

The **Earth** is nearly spherical. It has a layered structure that consists of a…

- thin **crust**
- **mantle**
- **core** (made of nickel and iron).

Rocks at the Earth's surface are continually being broken up, reformed and changed in an ongoing cycle of events, known as the **rock cycle**. The changes take a very long time.

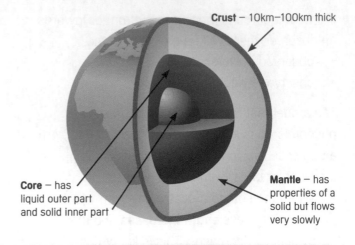

Crust – 10km–100km thick

Core – has liquid outer part and solid inner part

Mantle – has properties of a solid but flows very slowly

Tectonic Theory

At one time, scientists believed that features on the Earth's surface were caused by shrinkage of the crust when the Earth cooled down, following its formation.

But, as scientists have found out more about the Earth, this **theory** has now been rejected.

Evidence showed scientists that the east coast of South America and the west coast of Africa have…

- **closely matching coastlines**
- **similar patterns of rocks**, which contain **fossils** of the same plants and animals, e.g. the Mesosaurus.

This evidence led Alfred Wegener to propose that South America and Africa had at one time been part of a single land mass.

He proposed that the movement of the crust was responsible for the separation of the land, i.e. **continental drift**. This is **tectonic theory**.

Unfortunately, Wegener couldn't explain how the crust moved. It took more than 50 years for scientists to discover this.

How the Earth Once Was

Laurasia

Gondwanaland

How South America and Africa Could Have Once Looked

North America

South America

Africa

How the Earth Looks Today

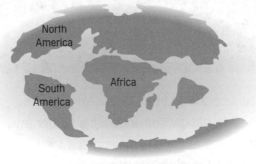

Key Words   **Theory • Evidence • Fossil**

Tectonic Plates

The Earth's lithosphere (the crust and the upper part of the mantle) is 'cracked' into **tectonic plates**.

Intense heat, released by radioactive decay deep in the Earth, creates convection currents in the mantle. These currents cause the tectonic plates to move apart very slowly by a few centimetres each year.

Convection in a gas or a liquid makes the matter rise as it is heated. As it gets further away from the heat source, it cools and sinks.

The same thing happens in the Earth:

1. Hot semi-molten mantle rises until it meets the crust, which can move.
2. The mantle begins to cool, then sinks down where the convections current starts to fall.
3. The land masses that ride on the moving mantle are known as plates. The plates move slowly.

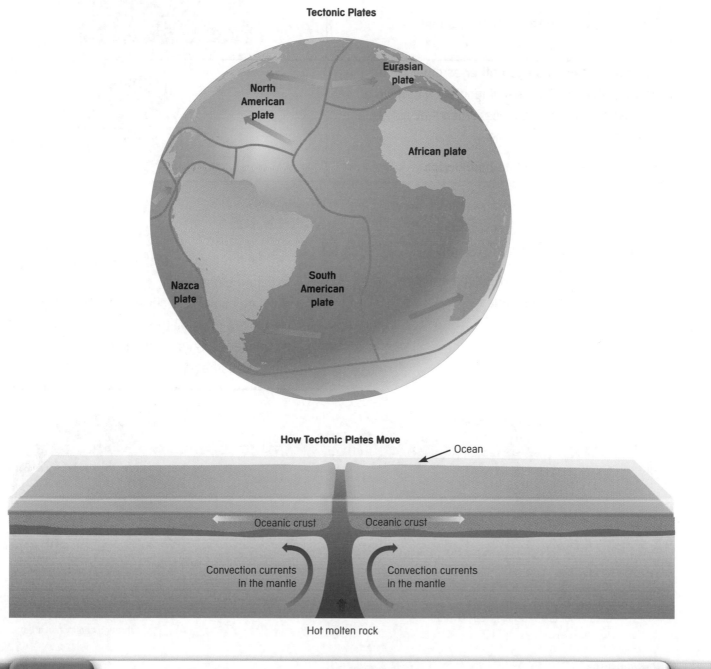

Tectonic Plates

Eurasian plate

North American plate

African plate

Nazca plate

South American plate

How Tectonic Plates Move

Ocean

Oceanic crust — Oceanic crust

Convection currents in the mantle — Convection currents in the mantle

Hot molten rock

Tectonic Plate Movement

Tectonic plates can move in three ways:

Plates slide past each other causing huge stresses and strains to build up in the crust. The eventual 'release' of this energy results in an earthquake. These are **conservative plate boundaries**.	An earthquake will occur along the line where the two plates meet
Plates move away from each other at an oceanic ridge, causing fractures to occur. Molten rock rises to the surface and solidifies to form new ocean floor. These are **constructive plate boundaries**.	Ocean floor Constructive plate boundary Magma rising
Plates move towards each other as other plates move away from each other, resulting in the formation of mountains and / or volcanoes. When they collide, one plate is forced under the other These are **destructive plate boundaries**.	Destructive plate boundary

Sudden plate movement can sometimes have disastrous consequences, for example, **earthquakes**, **volcanoes** and **tsunamis**.

Earthquakes and volcanic eruptions are common on destructive plate boundaries.

Quick Test

1. List the three layers that make up the structure of the Earth.
2. How are convection currents produced in the Earth's mantle?
3. What effects can sudden plate movements have?

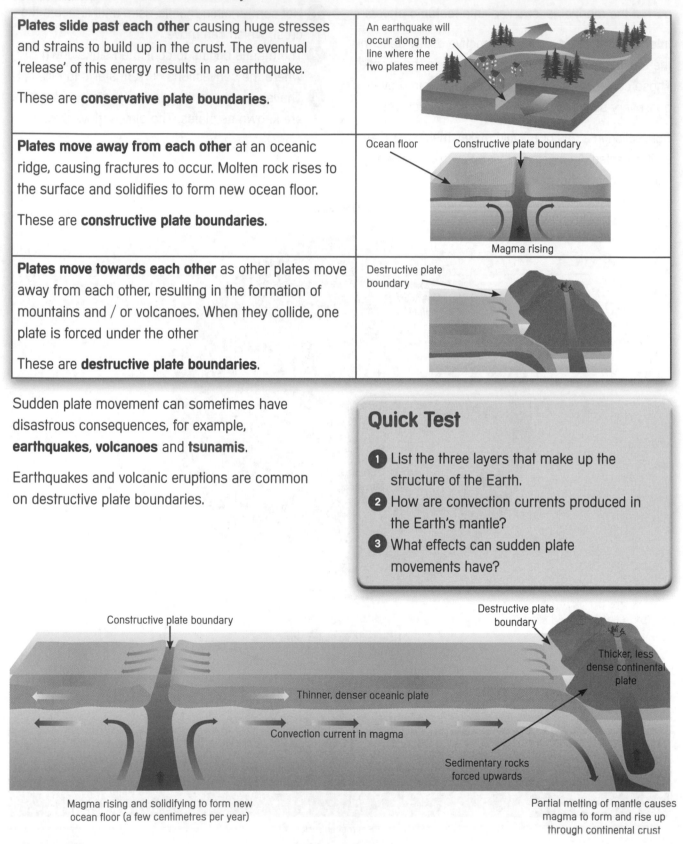

Constructive plate boundary

Destructive plate boundary

Thicker, less dense continental plate

Thinner, denser oceanic plate

Convection current in magma

Sedimentary rocks forced upwards

Magma rising and solidifying to form new ocean floor (a few centimetres per year)

Partial melting of mantle causes magma to form and rise up through continental crust

The Atmosphere

Since the formation of the Earth 4.6 billion years ago the atmosphere has changed a lot.

Time Scale	Condition of the Atmosphere	Key Factors and Events that Shaped the Atmosphere

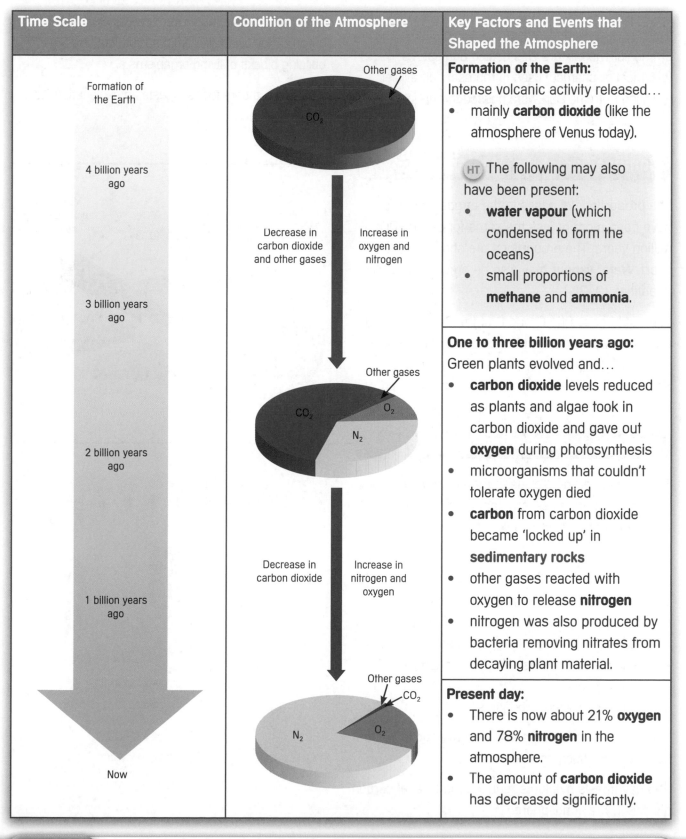

Time Scale

Formation of the Earth

4 billion years ago

3 billion years ago

2 billion years ago

1 billion years ago

Now

Condition of the Atmosphere

Other gases

CO₂

Decrease in carbon dioxide and other gases

Increase in oxygen and nitrogen

Other gases

CO₂   O₂   N₂

Decrease in carbon dioxide

Increase in nitrogen and oxygen

Other gases

CO₂

N₂   O₂

Key Factors and Events that Shaped the Atmosphere

Formation of the Earth:
Intense volcanic activity released...

- mainly **carbon dioxide** (like the atmosphere of Venus today).

(HT) The following may also have been present:

- **water vapour** (which condensed to form the oceans)
- small proportions of **methane** and **ammonia**.

One to three billion years ago:
Green plants evolved and...

- **carbon dioxide** levels reduced as plants and algae took in carbon dioxide and gave out **oxygen** during photosynthesis
- microorganisms that couldn't tolerate oxygen died
- **carbon** from carbon dioxide became 'locked up' in **sedimentary rocks**
- other gases reacted with oxygen to release **nitrogen**
- nitrogen was also produced by bacteria removing nitrates from decaying plant material.

Present day:

- There is now about 21% **oxygen** and 78% **nitrogen** in the atmosphere.
- The amount of **carbon dioxide** has decreased significantly.

U1 Our Changing Planet

HT The Origin of Life

Two scientists, **Miller** and **Urey**, tried to test one possible theory for how life on Earth began. They mixed together the chemicals thought to be present in the Earth's early atmosphere – water, methane and ammonia. The mixture was heated and sparks (electrical discharges) used to represent ultraviolet radiation from the Sun were passed through it.

The mixture was cooled and the cycle was repeated many times.

After many cycles the mixture contained simple organic molecules, like amino acids, that are the building blocks of living organisms.

This is one theory that suggests an explanation for how life began.

Composition of the Atmosphere

The proportions of gases in the atmosphere have been more or less the same for about 200 million years. The proportions are shown in the pie chart. **Water vapour** may also be present in varying quantities (0–3%).

The **noble gases** (in Group 0 of the Periodic Table) are all chemically unreactive gases. They are used in filament lamps and electric discharge tubes. **Helium** is much less dense than air and is used in balloons.

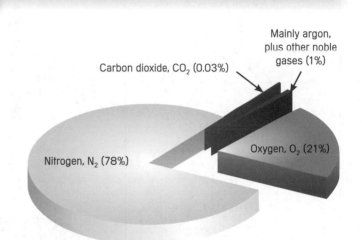

Mainly argon, plus other noble gases (1%)

Carbon dioxide, CO_2 (0.03%)

Oxygen, O_2 (21%)

Nitrogen, N_2 (78%)

Changes to the Atmosphere

The level of carbon dioxide in the atmosphere today is increasing due to several factors including:

- **The burning of fossil fuels** – burning carbon, which has been 'locked up' in **fossil fuels** for millions of years, releases carbon dioxide into the atmosphere.

The level of carbon dioxide in the atmosphere is reduced by the reaction between carbon dioxide and sea water. This reaction produces…

- insoluble carbonates, which are deposited as sediment
- soluble hydrogencarbonates.

The carbonates form the **sedimentary rocks** in the Earth's crust, such as limestone.

Photosynthesis by plants reduces the level of carbon dioxide in the atmosphere.

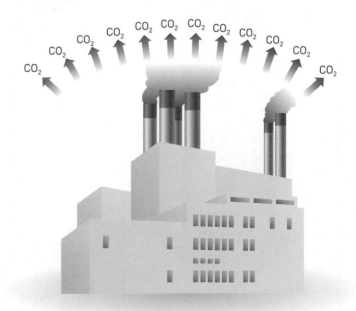

The Greenhouse Effect

Rays from the Sun reach Earth and are reflected back towards the atmosphere

CO_2 and CH_4 in the atmosphere absorb some of the energy and radiate it back to Earth. Carbon dioxide and methane allow short wave radiation to pass through the atmosphere to the Earth's surface ➤, but absorb the outgoing long wave radiation from the Earth ➤.

Some gases in the atmosphere, such as **methane** and carbon dioxide, prevent some heat 'escaping' from the Earth's surface into space. This keeps the temperature on Earth stable and warm enough to support life. This is known as the **greenhouse effect**.

The levels of these gases are slowly rising because of several factors:

- Increases in the numbers of cattle, sheep and rice fields mean more methane (CH_4) is being released into the atmosphere.
- The burning of chopped-down wood and fossil fuels, and some industrial processes means more carbon dioxide is being released into the atmosphere.

The increase of these greenhouse gases means that more heat is radiated back to Earth. This is causing **global warming**.

A rise in global temperature by only a few degrees Celsius could lead to…

- substantial climate changes
- a rise in sea level.

Quick Test

1. Since the formation of the Earth, name one gas that has decreased and two gases that have increased in the atmosphere.
2. Name one activity taking place now that is increasing the level of carbon dioxide in the atmosphere.
3. Describe two possible effects of a rise in global temperature.

Atoms and Elements

All substances are made of **atoms** (very small particles). Each atom has a small central **nucleus** made up of **protons** and **neutrons**. The nucleus is surrounded by orbiting **electrons**.

A substance that contains only one sort of atom is called an **element**. There are over 100 different elements.

The atoms of each element are represented by a different **chemical symbol**.

For example...
- sodium = Na
- carbon = C
- iron = Fe

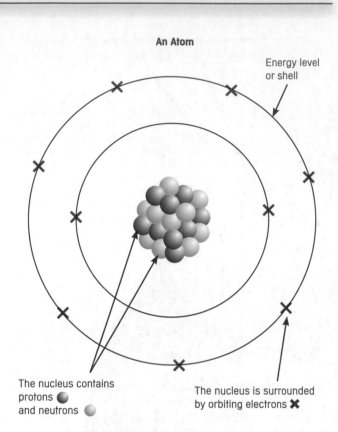

An Atom

Energy level or shell

The nucleus contains protons ● and neutrons ○

The nucleus is surrounded by orbiting electrons ✖

Mass Number and Atomic Number

Atoms of an element can be described using their **mass number** and **atomic number**.

The **mass number** is the total number of **protons** and **neutrons** in the nucleus of the atom.

The **atomic (proton) number** is the number of protons in the nucleus of the atom.

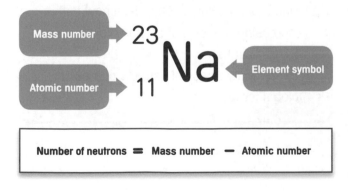

Mass number → 23
Atomic number → 11
Na
← Element symbol

Number of neutrons = Mass number − Atomic number

The number of **protons** in an atom is **equal** to the number of **electrons**, so an atom has no **overall charge**.

Hydrogen

$^{1}_{1}H$

1 proton
1 electron
0 neutrons

Oxygen

$^{16}_{8}O$

8 protons
8 electrons
8 neutrons

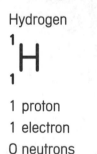

Atomic Particle	Relative Mass	Charge
Proton ●	1	+
Neutron ○	1	0
Electron ✖	Very small (negligible)	−

The Periodic Table

Elements are arranged in the **Periodic Table**. The **groups** in the Periodic Table contain elements that have similar properties.

For example, all Group 1 elements (the alkali metals) react vigorously with water to produce an alkaline solution and hydrogen gas.

You will need to know the electronic structure of the first 20 elements in the Periodic Table. See page 11 for a list of these elements and page 96 for their electronic structure.

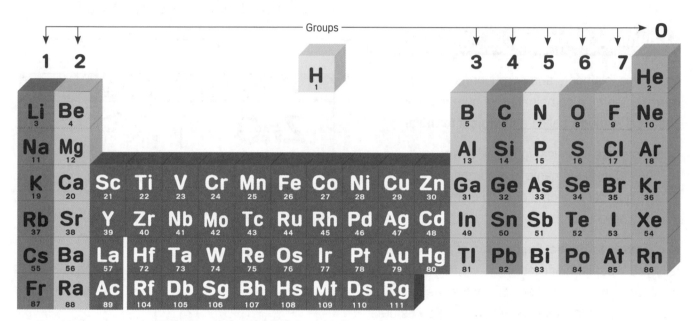

See the inner back cover for a full version of the Periodic Table

Electron Configuration and Structure

Electrons in an atom occupy the lowest available **energy level** (i.e. the innermost available shell):
*   The **first** level can only contain a **maximum** of **2 electrons**.
*   The energy levels after this can each hold more **electrons**.

The **electron configuration** tells us how the electrons are arranged around the nucleus in **energy levels** or **shells**.

It is written as a series of numbers, for example...
*   oxygen is 2,6
*   aluminium is 2,8,3

The **Periodic Table** groups the elements in terms of **electronic structure**.

Elements in the **same group** have the same number of **electrons** in their **highest energy level** (**outer shell**). They also have **similar** chemical **properties**.

A particular energy level is gradually filled with electrons from **left to right**, across each **period**. A period is a horizontal row of elements in the Periodic Table.

Mixtures and Compounds

A **mixture** consists of two or more elements or compounds that **aren't chemically combined**. For example, air is a mixture of gases.

When elements are chemically combined, the atoms are held together by **chemical bonds** (i.e. they're not just mixed together).

Atoms of two or more elements can be chemically combined to form **compounds**. For example, carbon dioxide, water and ammonia are all compounds.

When elements react, the atoms can form chemical bonds by…
- **sharing** electrons (covalent bond)
- **giving** or **taking** electrons (ionic bond).

Chemical Formulae

Molecules, elements and compounds can be represented by a combination of numbers and chemical symbols called a **chemical formula**.

Scientists use chemical formulae to show…
- the different elements in a compound
- the number of atoms of each element in the compound.

In **chemical formulae**, the position of the numbers tells you what is multiplied:
- A small number that sits below the line multiplies only the symbol that comes immediately before it.
- A number that is the same size as the letters multiplies all the symbols that come after it.

For example…
- H_2O means $(2 \times H) + (1 \times O)$
- $2NaOH$ means $2 \times (NaOH)$ or $2 \times (Na + O + H)$.

The number of capital letters equals the number of elements, e.g. **ZnO** = 2 elements.

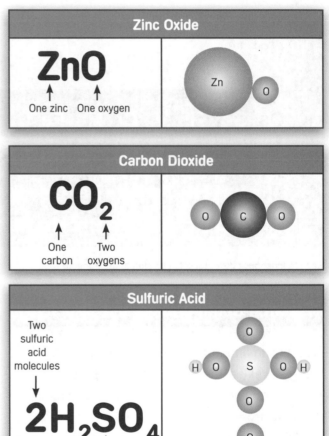

Zinc Oxide

One zinc   One oxygen

Carbon Dioxide

One carbon   Two oxygens

Sulfuric Acid

Two sulfuric acid molecules

Each one has…

Two hydrogens   One sulfur   Four oxygens

Quick Test

1. Name the three types of particles found in an atom.
2. How can the number of neutrons in a nucleus be calculated?
3. When different elements react with each other, they form chemical bonds. Describe how.
4. What do chemical formulae show?

Ions

Ions are negative or positive particles. Ions are formed when atoms give or take electrons. When atoms give electrons, they form **positive ions**. When atoms take electrons, they form **negative ions**.

Isotopes

All atoms of a **particular element** have the **same number** of protons. Atoms of **different elements** have **different numbers** of protons.

Isotopes are atoms of the **same element** that have **different numbers of neutrons**.

Isotopes have the **same atomic number** but a **different mass number**.

For example, chlorine has two isotopes:

$^{35}_{17}\text{Cl}$ $^{37}_{17}\text{Cl}$

17 protons
17 electrons
18 neutrons (35 − 17)

17 protons
17 electrons
20 neutrons (37 − 17)

Relative Atomic Mass, A_r

The **relative atomic mass** (A_r) of an element is found on the Periodic Table. It is the larger number shown for each element.

Relative atomic mass → 16

O

oxygen

8

The **relative atomic mass** is the mass of a particular atom compared with a twelfth of the mass of a carbon atom (the ^{12}C isotope).

The A_r is an **average** value for all the **isotopes** of the element.

For example, by looking at the Periodic Table, you can see that...

- carbon is 12 times heavier than hydrogen, but is only half as heavy as magnesium
- magnesium is three quarters as heavy as sulfur
- sulfur is twice as heavy as oxygen.

Relative Formula Mass, M_r

The **relative formula mass** (M_r) of a compound is the relative atomic masses of all its elements added together.

To calculate M_r you need to know...
- the formula of the compound
- the A_r of all the atoms involved.

Example 1
Calculate the M_r of water, H_2O. (A_r: H = 1, O = 16)

Write the formula → H_2O

Substitute the A_r → (2 × 1) + 16

Calculate the M_r → 2 + 16 = 18

Example 2
Calculate the M_r of potassium carbonate, K_2CO_3. (A_r: K = 39, C = 12, O = 16)

Write the formula → K_2CO_3

Substitute the A_r → (39 × 2) + 12 + (16 × 3)

Calculate the M_r → 78 + 12 + 48 = 138

Useful Materials

Useful materials can be removed from the ground by mining or quarrying.

Materials like gold, sulfur, limestone and marble can be used straight from the ground without any chemical processing.

Gold

Gold is an unreactive metal that's obtained from the ground by mining. Gold is used…

- to make jewellery because it's shiny and it doesn't corrode
- to coat the electrical connectors on electronic components because it's a good conductor of electricity.

Sodium Chloride

Sodium chloride (common salt) is a compound of an alkali metal and a halogen. It's found in large quantities in the sea and in underground deposits.

Salt is removed from underground deposits by mining.

Limestone

Limestone is a sedimentary rock that is removed from the ground by quarrying. It's used…
- as a building material
- for producing slaked lime
- for making glass
- for making cement, mortar and concrete.

Marble

Marble is a metamorphic rock that is removed from the ground by quarrying. Marble is used to make statues, buildings, tiles and worktops.

Sulfur

Sulfur, a non metal element, can be removed from underground using hot water. The hot water melts the sulfur, which is then pumped up to the surface.

Sulfur is used to make sulfuric acid, which is used to make other useful chemicals, e.g. fertilisers.

Separating Salt from Rock Salt

Rock salt, an impure mixture containing sand, rock, chemical impurities and salt, is dissolved in water. Impurities, including magnesium and calcium compounds are removed from the **brine**.

Brine is a solution of sodium chloride dissolved in water.

This solution is heated to make the water evaporate. Salt crystals remain on the surface of the container.

In hot countries, the brine is placed in large, open containers. Heat from the Sun is used to evaporate the water, leaving salt in the containers.

Crude Oil

Crude oil as it is taken from the ground isn't very useful. It is a **mixture** of compounds, some of which are very useful.

The properties of the substances in a mixture remain unchanged. They can be separated by physical methods, such as **distillation**.

Most of the compounds in crude oil consist of molecules made up of only **carbon** and **hydrogen** atoms. These compounds are called **hydrocarbons**. **Hydrocarbon** molecules vary in size. This affects their properties and how they are used.

The larger the hydrocarbon chain (i.e. the more carbon and hydrogen atoms in a molecule)...

- the less easily it flows (it's more viscous)
- the higher its boiling point
- the less volatile it is
- the less easily it ignites.

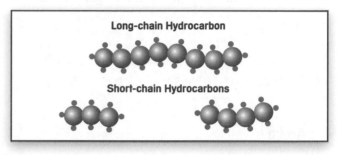

Long-chain Hydrocarbon

Short-chain Hydrocarbons

Fractional Distillation

Crude oil can be separated into different **fractions** (parts) by **fractional distillation**.

1 The crude oil is evaporated (by heating).

2 It is allowed to condense at a range of different temperatures.

3 The molecules collected at different temperatures are called **fractions**.

Each fraction contains hydrocarbon molecules with a similar number of carbon atoms. Most of the hydrocarbons obtained are **alkanes** (**saturated hydrocarbons**).

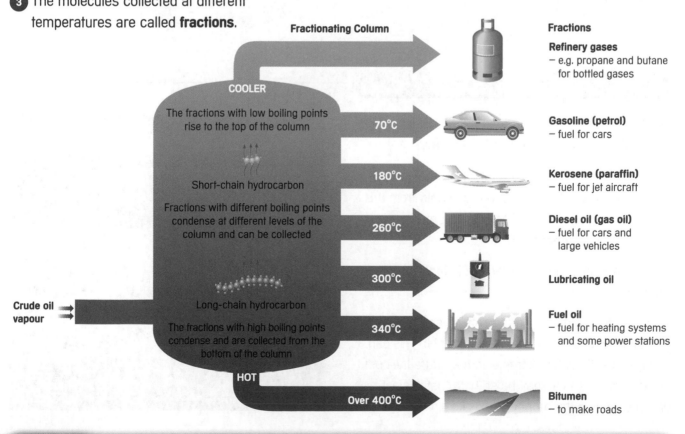

Fractionating Column

COOLER

The fractions with low boiling points rise to the top of the column

Short-chain hydrocarbon

Fractions with different boiling points condense at different levels of the column and can be collected

Long-chain hydrocarbon

The fractions with high boiling points condense and are collected from the bottom of the column

Crude oil vapour →

HOT

70°C

180°C

260°C

300°C

340°C

Over 400°C

Fractions

Refinery gases
— e.g. propane and butane for bottled gases

Gasoline (petrol)
— fuel for cars

Kerosene (paraffin)
— fuel for jet aircraft

Diesel oil (gas oil)
— fuel for cars and large vehicles

Lubricating oil

Fuel oil
— fuel for heating systems and some power stations

Bitumen
— to make roads

HT Air

Air (the atmosphere) is a mixture of gases with different boiling points. This means the mixture can be separated by **fractional distillation**.

Air is separated into useful new materials for industrial processes. Some of these useful materials are helium, argon and nitrogen.

Helium

Helium is less dense than air, so it can be used to inflate balloons as it allows the balloons to stay up in the air.

Argon

Argon is used in...
- electric filament bulbs to stop the hot filament catching fire
- electrical discharge tubes.

Nitrogen

Nitrogen gas is used to make ammonia gas, which is used to make fertilisers.

Ores

The Earth's crust contains many naturally occurring **elements** and **compounds** called **minerals**.

Some plants can remove metals from the ground, e.g. plants called 'Horsetails' remove gold from the soil. The plants can be collected and processed. This is called **phytomining**.

Some plants can remove metal pollutants from the ground.

A **metal ore** is a mineral that contains enough metal to make it economically viable to extract it. This can change over time, for example, in the future the cost of extracting a metal may be so high that it is not worth spending the money to extract it.

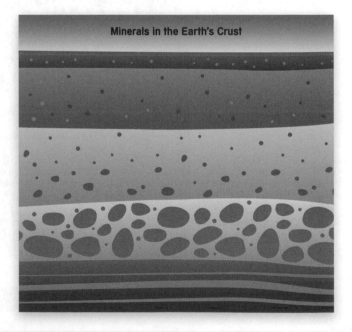

Minerals in the Earth's Crust

Extracting Metals from their Ores

The method of extraction for metals depends on **how reactive the metal is**.

Unreactive metals like gold exist naturally in the ground and can be obtained through mining and panning:

- Most metals are found as **metal oxides**, or as compounds that can be changed into a metal oxide. To extract a metal from its oxide, the oxygen must be removed by heating with another element in a **chemical reaction**. This process is called **reduction**.
- Metals that are **more reactive than carbon** can be extracted from their molten compounds by electrolysis, e.g. aluminium. A large amount of heat energy is used to melt the compounds of the metal. This is a costly process, so metals that have been extracted this way are **expensive**.
- Metals that are **less reactive than carbon** can be extracted from their oxides by heating with carbon, e.g. iron, lead.

Gold Panning

Iron

Iron oxide can be reduced in a blast furnace to produce **iron**. Molten iron obtained from a blast furnace contains roughly...

- 96% iron
- 4% carbon and other metals.

Because it is impure, the iron is very brittle with limited uses. To produce pure iron, all the **impurities** have to be removed.

The **atoms** in pure iron are arranged in layers, which can slide over each other easily. This makes pure iron soft and malleable. It can be easily shaped, but it's too soft for many practical uses.

The properties of iron can be changed by mixing it with small quantities of carbon and other metals to make **steel**. The majority of iron is converted into steel. Steel is an **alloy**.

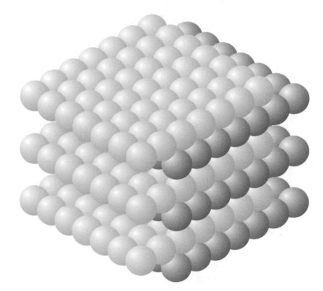

Pure Iron

Iron and Lead

Iron and lead can be extracted from their oxides by reduction.

Extracting Iron

Iron is extracted from iron oxide by heating it with coke and limestone. Coke contains a high proportion of carbon.

1 Coke (carbon) reacts with oxygen to make carbon dioxide.

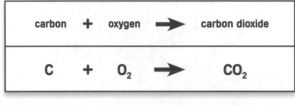

carbon	+	oxygen	→	carbon dioxide
C	+	O_2	→	CO_2

2 Carbon dioxide reacts with more coke to make carbon monoxide.

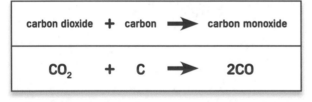

carbon dioxide	+	carbon	→	carbon monoxide
CO_2	+	C	→	$2CO$

3 The carbon monoxide reduces the iron oxide to iron.

iron oxide	+	carbon monoxide	→	iron	+	carbon dioxide
Fe_2O_3	+	$3CO$	→	$2Fe$	+	$3CO_2$

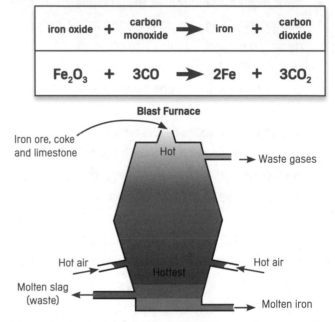

Limestone removes some impurities and releases carbon dioxide.

Extracting Lead

Lead oxide is heated with carbon or carbon monoxide, which reduces the lead oxide. The carbon or carbon monoxide is the **reducing agent**.

Using carbon:

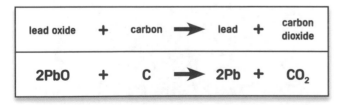

lead oxide	+	carbon	→	lead	+	carbon dioxide
$2PbO$	+	C	→	$2Pb$	+	CO_2

Using carbon monoxide:

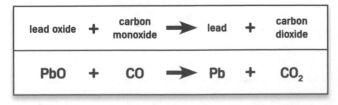

lead oxide	+	carbon monoxide	→	lead	+	carbon dioxide
PbO	+	CO	→	Pb	+	CO_2

Quick Test

1 List two ways of removing useful materials from the ground.
2 Name the process that can be used to separate a mixture of gases.
3 Why is aluminium expensive to extract from aluminium oxide?
4 Give the names of two reducing agents.
5 Write a word equation to represent the reaction between iron oxide and carbon monoxide.

Calculating Percentage Mass

The mass of a compound is its **relative formula mass**.

To calculate the **percentage mass** of an element in a compound, you need to know...

- the **formula** of the compound
- the **relative atomic mass** of all the atoms.

You can calculate the percentage mass by using this formula:

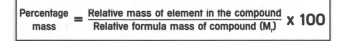

$$\text{Percentage mass} = \frac{\text{Relative mass of element in the compound}}{\text{Relative formula mass of compound } (M_r)} \times 100$$

Example 1

Calculate the percentage mass of magnesium in magnesium oxide, MgO.

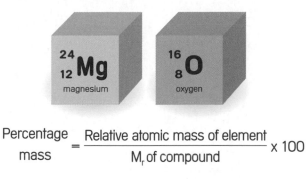

magnesium   oxygen

$$\text{Percentage mass} = \frac{\text{Relative atomic mass of element}}{M_r \text{ of compound}} \times 100$$

Relative atomic mass of magnesium = 24
Relative formula mass (M_r) of MgO =

$$24 + 16 = 40$$

A_r Mg A_r O M_r MgO

Percentage mass = $\frac{24}{40}$ x 100 = **60%**

Example 2

Calculate the percentage mass of potassium in potassium carbonate, K_2CO_3.

potassium   carbon   oxygen

Relative atomic mass of potassium = 39 x 2
Relative formula mass (M_r) of K_2CO_3 =

$$78 + 12 + 48 = 138$$

A_r K x 2 A_r C A_r O x 3 M_r K_2CO_3

Percentage mass = $\frac{78}{138}$ x 100 = **56.5%**

Empirical Formula of a Compound

The empirical formula of a compound is the **simplest formula** that represents the **composition** of the compound **by mass**.

Example
Find the simplest formula of an oxide of iron produced by reacting 1.12g of iron with 0.48g of oxygen (A_r Fe = 56, A_r O = 16).

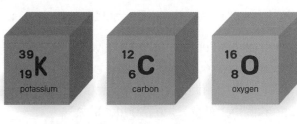

Identify the mass of the elements in the compound

Masses: Fe = 1.12, O = 0.48

Divide these masses by their relative atomic masses

$$Fe = \frac{1.12}{56} = 0.02 \qquad O = \frac{0.48}{16} = 0.03$$

Identify the ratio of atoms in the compound

Ratio =   x 100   0.02 : 0.03   x 100

2 : 3

Empirical formula = **Fe₂O₃**

Chemical Reactions

You can show what has happened during a **chemical reaction** by writing a **word equation**.

The **reactants** (i.e. the substances that react) are on the left hand side of the equation and the **products** (i.e. the new substances that are formed) are on the right hand side.

The total mass of the products of a chemical reaction is always equal to the total mass of the reactants. This is because **no atoms are lost or made**. The products of a chemical reaction are made up from exactly the same atoms as the reactants.

Chemical symbol equations must always be **balanced**. There must be the same number of atoms of each element on the reactant side of the equation as there is on the product side.

Number of atoms of each element on reactants side	=	Number of atoms of each element on products side

Example

	Reactants	→	Products
Word equation	sodium + water	→	sodium hydroxide + hydrogen
Symbol equation	2Na + 2H$_2$O	→	2NaOH + H$_2$

This means that...	2 atoms of sodium	and	2 molecules of water	produce	2 molecules of sodium hydroxide	and	1 molecule of hydrogen

Quick Test

1. What is an isotope?
2. What formula would you use to calculate the percentage mass of an element in a compound?

Calculating the Mass of Product

Example

Calculate how much calcium oxide can be produced from 50kg of calcium carbonate.
(Relative atomic masses: Ca = 40, C = 12, O = 16).

1. Write down the equation.
2. Work out the M_r of each substance.
3. Check that the total mass of reactants equals the total mass of the products. If they are not the same, check your work.
4. The question only mentions calcium oxide and calcium carbonate, so you can now ignore the carbon dioxide. You just need the ratio of mass of reactant to mass of product.
5. Use the ratio to calculate how much calcium oxide can be produced.

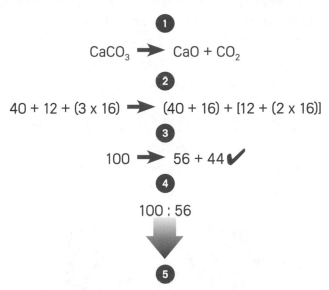

1

$$CaCO_3 \rightarrow CaO + CO_2$$

2

$$40 + 12 + (3 \times 16) \rightarrow (40 + 16) + [12 + (2 \times 16)]$$

3

$$100 \rightarrow 56 + 44 ✔$$

4

$$100 : 56$$

5

If 100kg of $CaCO_3$ produces 56kg of CaO,
then 1kg of $CaCO_3$ produces $\frac{56}{100}$ kg of CaO,
and 50kg of $CaCO_3$ produces $\frac{56}{100} \times 50$

= 28kg of CaO

Calculating the Mass of Reactant

Example

Calculate how much aluminium oxide is needed to produce 540 tonnes of aluminium.
(Relative atomic masses: Al = 27, O = 16).

1. Write down the equation.
2. Work out the M_r of each substance.
3. Check that the total mass of reactants equals the total mass of the products. If they are not the same, check your work.
4. The question only mentions aluminium oxide and aluminium, so you can now ignore the oxygen. You just need the ratio of mass of reactant to mass of product.
5. Use the ratio to calculate how much aluminium oxide is needed.

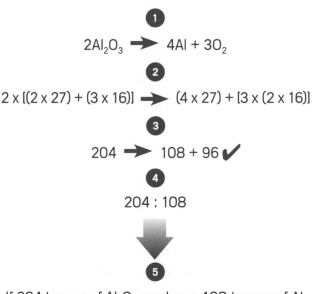

1

$$2Al_2O_3 \rightarrow 4Al + 3O_2$$

2

$$2 \times [(2 \times 27) + (3 \times 16)] \rightarrow (4 \times 27) + [3 \times (2 \times 16)]$$

3

$$204 \rightarrow 108 + 96 ✔$$

4

$$204 : 108$$

5

If 204 tonnes of Al_2O_3 produces 108 tonnes of Al,
then $\frac{204}{108}$ tonnes is needed to produce 1 tonne of Al,
and $\frac{204}{108} \times 540$ tonnes is needed to
produce 540 tonnes of Al

= 1020 tonnes of Al_2O_3

U1 Using Materials from Our Planet to Make Products

Yield

Atoms are **never lost or gained** in a chemical reaction. But, it isn't always possible to obtain the calculated amount of the product because…

- if the reaction is reversible, it might not go to completion
- some product could be lost when it is separated from the reaction mixture
- some of the reactants may react in different ways to the expected reaction.

The amount of product obtained is called the **yield**.

The **percentage yield** can be calculated by comparing…

- the actual yield obtained from a reaction
- the maximum theoretical yield.

$$\text{Percentage yield} = \frac{\text{Yield from reaction}}{\text{Maximum theoretical yield}} \times 100$$

Example

50kg of calcium carbonate ($CaCO_3$) is expected to produce 28kg of calcium oxide (CaO).

A company heats 50kg of calcium carbonate in a kiln and obtains 22kg of calcium oxide. Calculate the percentage yield.

Percentage yield = $\frac{22}{28}$ x 100 = **78.6%**

Calculating Atom Economy

Chemical reactions often produce more than one **product**. However, not all of these products are '**useful**'. A useful product is one that can be used in industry.

Atom economy (atom utilisation) is a measure of the **amount of starting materials** (reactants) that end up as **useful products**.

$$\text{Atom economy} = \frac{M_r \text{ of useful products}}{M_r \text{ of reactant}} \times 100$$

In industry, it's important to use reactions that have a **high atom economy** in order to…

- make the reaction economical
- ensure sustainable development.

Quick Test

1. Why is it important to use reactions that have high atom economy?
2. What formula would you use to calculate the percentage yield of a product from a reaction?

Example

calcium carbonate → calcium oxide + carbon dioxide

The products of this reaction are calcium oxide (useful) and carbon dioxide (waste).

Calculate the atom economy of this reaction.

First calculate the M_r of the reactants and products.

$CaCO_3$ → CaO + CO_2

M_r = 40 + 12 + (3 x 16) 40 + 16 12 + (2 x 16)
= 100 = 56 = 44

Atom economy = $\frac{56}{100}$ x 100
= **56%**

HT Writing Balanced Equations

The following steps tell you how to write a balanced equation.

1. Write a word equation for the chemical reaction.
2. Substitute in formulae for the elements or compounds.
3. Balance the equation by adding numbers in front of the reactants and / or products.
4. Write down the balanced symbol equation.

Reactants			→	Products
1 Write a word equation	magnesium	+ oxygen	→	magnesium oxide
2 Substitute in formulae	Mg	+ O_2	→	MgO

3 Balance the equation

- There are two **O**s on the reactant side, but only one **O** on the product side. We need to add another **MgO** to the product side to balance the **O**s
- We now need to add another **Mg** on the reactant side to balance the **Mg**s
- There are two magnesium atoms and two oxygen atoms on each side – **it is balanced**.

4 Write a balanced symbol equation	2Mg	+ O_2	→	2MgO

Reactants			→	Products
1 Write a word equation	nitrogen	+ hydrogen	→	ammonia
2 Substitute in formulae	N_2	+ H_2	→	NH_3

3 Balance the equation

- There are two **N**s on the reactant side, but only one **N** on the product side. We need to add another **NH₃** to the product side to balance the **N**s
- We now need to add two more **H₂**s on the reactant side to balance the **H₂**s
- There are two nitrogen atoms and six hydrogen atoms on each side – **it is balanced**.

4 Write a balanced symbol equation	N_2	+ $3H_2$	→	$2NH_3$

U1 Life on Our Planet

Classification

There is a wide variety of life that can be sorted into large groups called **kingdoms**. Kingdoms are sorted into groups by comparing similar characteristics.

Each group is then sorted into smaller groups by comparing similar features.

The number of different **species** present in each group reduces until only one species is left.

Scientists all over the world use the same system to classify living things into groups. Animal and plant species can be identified and named wherever they are found using this system.

Kingdoms

- **Plants**
  - They have cellulose cell walls
  - They photosynthesise

- **Animals**
  - They have a nervous system

- **Fungi**
  - They do not contain chlorophyll

- **Protoctists (mostly single cells)**
  - They do not have specialised cells

- **Prokaryotes (bacteria)**
  - They do not have a nucleus

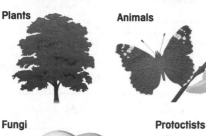

Plants   Animals

Fungi   Protoctists

Vertebrates

- **Fish**, e.g. John Dory fish
  - cold-blooded; have gills; lay eggs in water; have scales

- **Amphibians**, e.g. frog
  - cold-blooded; adults have lungs (young have gills); lay eggs in water; have smooth skin

- **Reptiles**, e.g. snake
  - cold-blooded; have lungs; lay eggs; have scaly skin

- **Birds**, e.g. skylark
  - warm-blooded; have lungs; lay eggs; have feathers

- **Mammals**, e.g. sheep
  - warm-blooded; have lungs; give birth to live offspring; have hair or fur on skin

Populations and Communities

A **population** is the total number of individuals of the same species that live in a certain area.

A **community** is the total number of organisms of all species in a particular area.

Competition

In order to **survive**, organisms use materials from their surroundings and from the other living organisms there.

Organisms **compete** with each other for **food**, **water** and **space**.

Plants compete with each other for...
- light
- water
- nutrients from the soil.

Animals compete with each other for...
- food
- water
- mates
- territory
- shelter.

When organisms compete, those that are **better adapted** to their environment are more successful and usually exist in larger numbers. This often leads to the exclusion of other competing organisms.

Adaptations

Adaptations are features that make an organism well-suited to its environment. Adaptations increase an organism's chance of **survival**.

Animals and plants are adapted for survival in the conditions of their environment. For example, the polar bear is well-suited to its habitat in the Arctic:
- It has a white coat so it is camouflaged.
- It has a lot of insulating fat beneath its skin.
- It has large feet to spread its weight on the ice.
- It has a small surface area to volume ratio.

The following are some more adaptations:
- Some plants that live in dry conditions have **water storage tissues** and **extensive root systems** to absorb as much water as possible.
- Plants that live in the shade have a **large surface area** of leaf to absorb as much light as possible.
- Some animals that live in dry conditions have adaptations that allow them to store water.

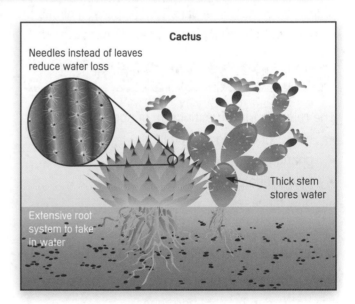

Some microbes (**extremophiles**) live in very extreme conditions. For example...
- very cold environments, e.g. the Arctic
- very hot environments, e.g. volcanic vents
- very dry environments, e.g. deserts
- severe chemical environments, e.g. sea water.

The Theory of Evolution

The **theory of evolution** states that all living things that exist today, and many more that are now **extinct**, **evolved** from **simple life forms** that first developed three billion years ago.

Studying the similarities and differences between species can help us to understand evolution. **Fossils** are the **remains of plants or animals** from many years ago that are found in rock. Fossils provide **evidence** of how organisms have changed over time.

Evolution by Natural Selection

Evolution is the change in the characteristics in a population over many generations. It may result in the formation of a new species. The members of the new species are **better adapted** to their environment.

For example, most peppered moths were originally pale in colour, which meant they were **camouflaged** against the bark of silver birch trees and **predators** found it hard to see (and eat) them.

But during the Industrial Revolution the air became polluted and silver birch trees turned black with soot. The pale coloured moths were now more easily seen and eaten by predators. **Natural selection** led to a greater proportion of the darker variety of peppered moth as they were better camouflaged against the darker trees.

1. Individual organisms within a species show **variation**.
2. Individuals better adapted to their environment are more likely to survive, breed successfully and produce offspring. This is **survival of the fittest**.
3. These survivors will pass on their genes to their offspring, resulting in an improved organism being **evolved** through **natural selection**.

Where new forms of a gene result from **mutation**, there may be a more rapid change in a species.

Peppered Moth   **Dark Peppered Moth**

Extinction of Species

The reasons for the **extinction** of species include…
- new / increased competition
- changes in the environment
- new predators
- new diseases.

Quick Test

1. What do scientists look at when they sort living things into groups?
2. What do plants compete for?
3. Name two ways in which plants are adapted so they survive in dry conditions.

Factors that Affect Plant Growth

Plant hormones control the growth of plants. **Auxins** are one group of plant hormones.

Plant shoots grow towards **light**. This response is called **phototropism**. Plants produce flowers and fruits at specific times of year depending on **day length**.

Plant shoots and roots respond to **gravity**. This is called **gravitropism**. Shoots grow away from gravity. Plant roots grow towards gravity.

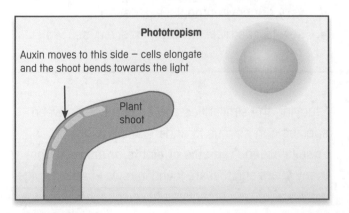

Phototropism

Auxin moves to this side – cells elongate and the shoot bends towards the light

Plant shoot

Gravitropism

Root

Auxin moves to this side – growth is inhibited so root grows downwards

Food Chains

Energy from the Sun is the source of **energy** for all **communities** of living organisms. In green plants, photosynthesis uses a **small fraction** of the solar energy that reaches them.

This energy is stored in the substances that make up the cells of the plant. It can then be passed on to organisms that eat the plant. This transfer of energy can be represented by a **food chain**.

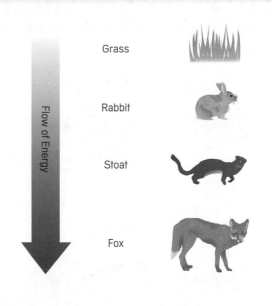

Flow of Energy

Grass

Rabbit

Stoat

Fox

Pyramids of Biomass

Biomass is the mass of living materials. In a food chain, the biomass at each stage is **less** than it was at the previous stage.

The biomass at each stage can be drawn to scale and shown as a **pyramid of biomass**.

Biomass and energy are '**lost**' at every stage of a food chain. This is due to...
- materials and energy being lost in an organism's faeces (waste)
- energy released through **respiration** being used for movement and heat energy.

Mammals and birds, in particular, lose a lot of energy. They must keep their body temperature constant, which is often higher than their surroundings.

In the biomass pyramid:
- Only a small amount of the Sun's energy is used by the producers.
- Rabbits respire and release waste products. They pass on $\frac{1}{10}$ of the energy they get from the grass.
- Stoats respire and release waste products. They pass on $\frac{1}{10}$ of the energy they get from the rabbits.
- The fox gets the last bit of energy and biomass.

Since the loss of energy and biomass is due mainly to **heat loss**, **waste** and **movement**, the efficiency of **food production** can be improved by...
- reducing the number of stages in a food chain
- limiting an animal's movement
- controlling the temperature of the animal's surroundings.

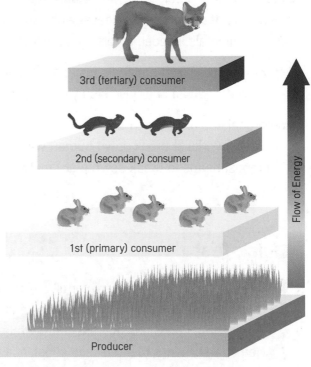

3rd (tertiary) consumer

2nd (secondary) consumer

1st (primary) consumer

Producer

Flow of Energy

Calculating Energy Transfer

The percentage of energy transfer can be calculated at each stage in a food chain.

For example, a food chain looks like this:

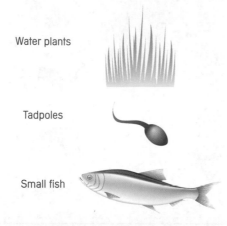

Water plants

Tadpoles

Small fish

The tadpoles get 108 000kJ of energy from the water plants. The small fish get 18 000kJ of energy from the tadpoles.

$$\text{Percentage energy transferred} = \frac{\text{Energy given to small fish}}{\text{Energy given to tadpoles}}$$

$$= \frac{18\,000}{108\,000} \times 100$$

$$= 16\%$$

Recycling the Materials of Life

Living things (organisms) **remove materials** from the **environment** for growth and other processes.

When the organisms die or excrete waste, the materials are **returned** to the environment.

Microorganisms break down waste and dead bodies. This **decay process** releases substances, including minerals, needed by plants for growth.

Microorganisms digest materials faster in conditions that are **warm**, **moist** and have plenty of **oxygen**.

Microorganisms are used in…
* sewage works to break down **human waste**
* compost heaps to break down **plant material**.

Eating

Waste

Death

Broken down by microorganisms

Absorption

Quick Test

1. What is the source of energy for all communities of living organisms?
2. Give two reasons why biomass is lost at every stage of a food chain.
3. Name three conditions that affect growth in microorganisms.
4. Name the process that releases substances used for plant growth from waste or dead bodies.

The Carbon Cycle

In a stable community, two processes are **balanced**:

- The **removal** of materials from the environment.
- The **return** of materials to the environment.

This constant recycling of carbon-containing material is called the **carbon cycle**.

1 Carbon dioxide, CO_2, is **removed** from the atmosphere by green plants for **photosynthesis**. Some CO_2 is **returned** to the atmosphere by plants during **respiration**.

2 The carbon obtained by photosynthesis is used by **plants** to make carbohydrates, fats and proteins. The plants are eaten by **animals**. Some of the carbon becomes fats, proteins and carbohydrates in the animals.

3 Animals **respire**, releasing CO_2 into the **atmosphere**.

4 Plants and animals **die**. Other animals and microorganisms feed on their bodies, causing them to **break down**.

5 **Detritus** feeders and microorganisms respire, releasing CO_2 into the atmosphere.

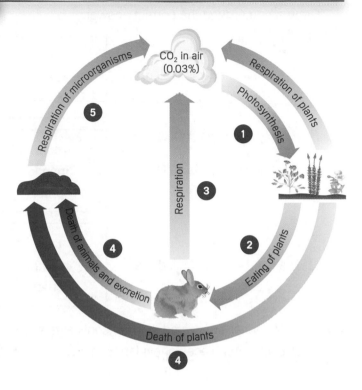

Storing Carbon

Carbon is stored in fossil fuels but is released as carbon dioxide when fossil fuels are burned.

Carbon dioxide dissolves in water and is used to make calcium carbonate (limestone).

This is used by animals to make shells and bones. This process takes place in warm, tropical, shallow oceans where limestone-producing organisms live.

Carbon is stored in the shells and bones of marine animals and plankton that collect on the ocean floor, e.g. corals. This process takes a long time.

The more warm, tropical, shallow oceans there are, the greater the amount of limestone deposited.

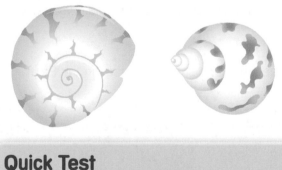

Quick Test

1 Describe the two processes that are balanced in the carbon cycle.

2 Name the process that takes place constantly in plants and animals that releases carbon dioxide into the air.

3 Describe two ways in which carbon is stored.

1 Scientists have discovered that the Earth has three layers.   **(3 marks)**
Draw a ring around the correct answer.

a) The outside layer is called the

> core
> mantle
> crust

b) The middle layer is called the

> core
> mantle
> crust

c) The inner layer is called the

> core
> mantle
> crust

2 Alfred Wegener suggested that the continents had once been a single land mass.
a) Describe two pieces of evidence that support this theory.   **(2 marks)**

....

....

b) The Earth's lithosphere is cracked into tectonic plates. Describe how tectonic plates can move.   **(3 marks)**

....

....

....

c) The movement of tectonic plates can have disastrous effects. Name two of these effects.   **(2 marks)**

....

3 The pie charts show the gases in the Earth's early atmosphere and the Earth's atmosphere today.

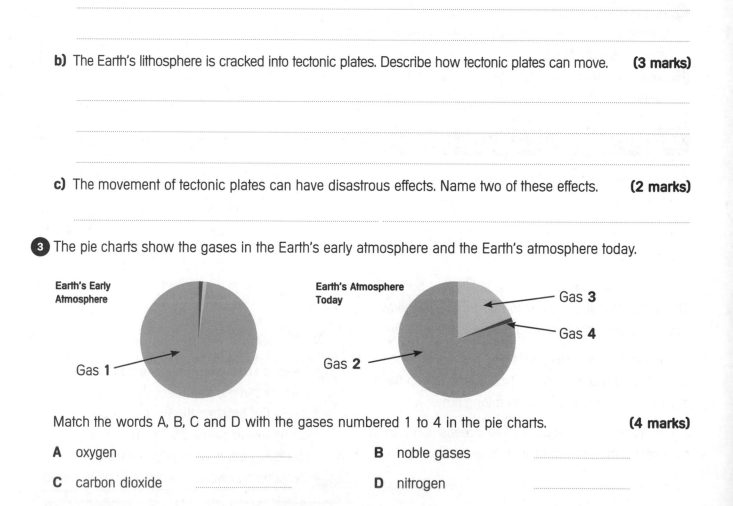

Earth's Early Atmosphere

Gas **1**

Earth's Atmosphere Today

Gas **2**

Gas **3**

Gas **4**

Match the words A, B, C and D with the gases numbered 1 to 4 in the pie charts.   **(4 marks)**

A oxygen ....   **B** noble gases ....

C carbon dioxide ....   **D** nitrogen ....

4 Darwin suggested his theory about evolution nearly 200 years ago.

a) Explain the meaning of the word 'evolution'. **(2 marks)**

....

....

b) Fossils have been used to support the theory of evolution.

i) What are fossils? **(2 marks)**

....

....

ii) How can fossils be used to support the theory of evolution? **(1 mark)**

....

....

5 There are many factors that affect the survival of a species.
Choose the correct words from the box to complete the sentences below. **(4 marks)**

environment	**population**	**species**	**competition**	**predators**	**prey**	**disease**

Extinction may happen when there is increased .... between organisms or if the ....

changes too quickly. The introduction of new .... or a new .... can also cause

species to become extinct.

6 Sally noticed a blackbird, sitting in an oak tree, eating ladybirds. When she looked closer she saw that there were greenfly on the leaves of the tree, and the greenfly were being eaten by the ladybirds.

The diagram below represents the pyramid of biomass relating to the food chain above. Write the name of each organism in the correct place at the side of each stage. **(4 marks)**

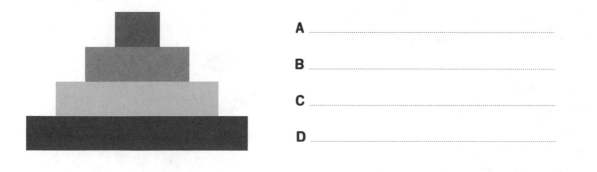

A ....

B ....

C ....

D ....

The Parts of the Nervous System

Your nervous system allows you to…
- **react** to your surroundings
- **coordinate** your behaviour.

It consists of…
- your **brain**
- your **spinal cord**
- **paired spinal nerves**
- **receptors**.

Information from **receptors** in your sense organs passes along **neurones** (nerve cells) to your brain. Your brain then coordinates your response by sending instructions to **effectors**.

Brain

Spinal cord

Spinal nerves

Nervous System

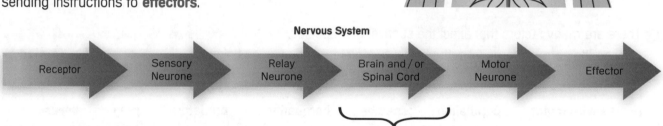

Receptor → Sensory Neurone → Relay Neurone → Brain and / or Spinal Cord → Motor Neurone → Effector

The Central Nervous System (CNS)

The Three Types of Neurone

Neurones are **specially adapted cells** that carry **electrical signals**, e.g. nerve impulses:
- They are **elongated** (long and thin) to **make connections** between different parts of your body.
- They have **branched endings**; for example, a **single neurone** can act on **lots of muscle fibres** at the same time.

Motor Neurone — Impulse travels away from cell body

Sensory Neurone — Impulse travels towards cell body

Relay Neurone — Impulse travels first towards and then away from cell body

Connections Between Neurones

The neurones don't touch each other. There is a tiny gap between them called a **synapse**.

1. An electrical **impulse** reaches the synapse through neurone A.
2. A **chemical transmitter** is released.
3. The transmitter activates **receptors** on neurone B.
4. An electrical **impulse** is generated in neurone B.

The **chemical transmitter** is **destroyed**.

Neurone A

Cell body of neurone B

Reflex Action

Reflex actions are designed to prevent your body from being harmed. For example, if you touch something hot, your hand automatically jerks away from it. If this was a conscious action, i.e. you had to think about the best way to respond, the process would be much slower and your hand would get burned.

Reflex actions are automatic and quick. They speed up your response time by by-passing your brain. Your spinal cord acts as the coordinator. The spinal cord passes an impulse directly…

- from a **sensory neurone**…
- through a **relay neurone**…
- to a **motor neurone**.

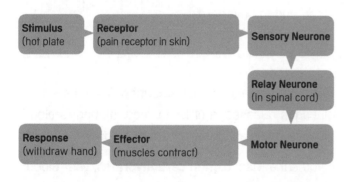

Stimulus (hot plate	Receptor (pain receptor in skin)		Sensory Neurone
			Relay Neurone (in spinal cord)
Response (withdraw hand)	Effector (muscles contract)		Motor Neurone

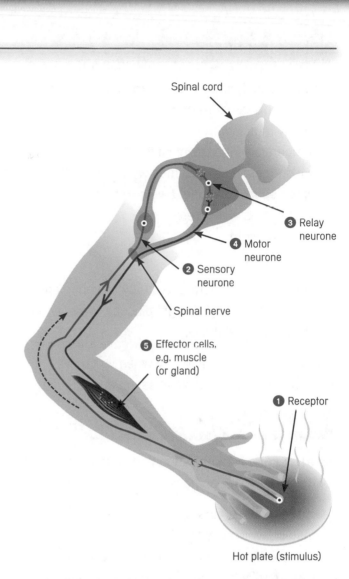

Spinal cord

❸ Relay neurone

❹ Motor neurone

❷ Sensory neurone

Spinal nerve

❺ Effector cells, e.g. muscle (or gland)

❶ Receptor

Hot plate (stimulus)

Types of Receptor

The **receptors** in your sense organs work to detect **stimuli** (changes in your environment). Different stimuli are detected by different receptors:

- **Light** – receptors in your eyes.
- **Change of position** – receptors in your ears (for balance).
- **Taste** – receptors on your tongue.
- **Smell** – receptors in your nose.
- **Touch**, **pressure**, **pain** and **temperature** – receptors in your skin.
- **Sound** – receptors in your ears.

We hear sounds because **longitudinal waves** travel from vibrating objects to our ears. Our hearing range is 20–20 000Hz.

The information about changes in your environment follows this pathway through your body to produce a response:

Nerve Pathway

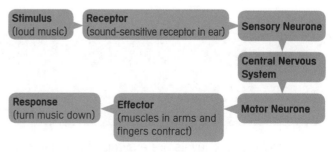

Stimulus (loud music)	Receptor (sound-sensitive receptor in ear)	Sensory Neurone
		Central Nervous System
Response (turn music down)	Effector (muscles in arms and fingers contract)	Motor Neurone

This is all coordinated by your **central nervous system**, which receives information (in the form of impulses) via the spinal nerves.

Internal Conditions

Humans keep their internal environment relatively constant. This is called **homeostasis**. This means your body must control…

- temperature – to maintain the temperature at which most body enzymes work best (37°C)
- **blood sugar** (**glucose**) **levels** – to constantly supply cells with energy
- **water content**
- **ion content**.

Your body raises its temperature by shivering and lowers it by sweating. You get glucose, water and ions by eating and drinking.

Glucose is used up as energy when you move. Water and ions are lost by sweating and through your kidneys in urine. Water is also lost through your lungs when you breathe out.

Hormones

Many processes within your body are coordinated by **hormones**. Hormones are chemicals produced by **glands**, and transported to their target organs by the bloodstream.

Hormones regulate the functions of many organs and cells. Women produce hormones that cause…

- eggs to be released from the **ovaries**
- changes in the thickness of the **womb** lining.

These hormones are produced by the **pituitary gland** and the ovaries.

Glucose

Blood glucose concentration is monitored and controlled by your pancreas.

The pancreas secretes two **hormones**:
- **Insulin**, which causes the liver to change glucose in your blood to insoluble glycogen.
- **Glucagon**, which causes the liver to change insoluble glycogen to glucose.

The pancreas adjusts the amount of insulin and glucagon released in order to keep the body's blood glucose levels as close to normal as possible.

If you eat a meal high in **carbohydrates**, your blood glucose concentration will go up, so your pancreas produces insulin.

If you exercise, your blood glucose concentration will go down, so your pancreas produces glucagon.

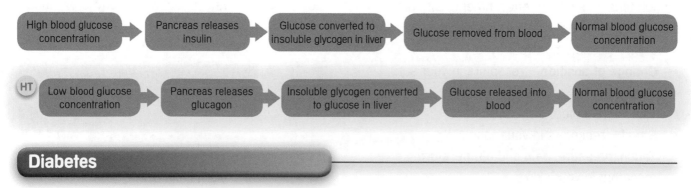

High blood glucose concentration → Pancreas releases insulin → Glucose converted to insoluble glycogen in liver → Glucose removed from blood → Normal blood glucose concentration

HT — Low blood glucose concentration → Pancreas releases glucagon → Insoluble glycogen converted to glucose in liver → Glucose released into blood → Normal blood glucose concentration

Diabetes

High blood glucose levels are a symptom of diabetes.
- Type 1 diabetes (insulin-dependent diabetes) is controlled by insulin doses that may be injected.
- Type 2 diabetes can be controlled by exercise and a change of diet.

HT Negative Feedback

Negative feedback helps to maintain a constant internal environment. It helps to bring conditions back to normal.

An example of negative feedback in action is the way the body maintains normal blood glucose levels. If there is a high blood glucose concentration insulin is produced, but once the concentration has returned to normal no more insulin will be produced.

Body Temperature

Your body temperature is **monitored** and **controlled** by the **thermoregulatory centre** in the brain. This centre has receptors that are sensitive to the temperature of blood flowing through them.

Temperature receptors in the skin also provide information about skin temperature. Sweat helps to cool your body when it is **hot**. When you sweat, water is **lost** from the body, so more water has to be **taken** in as food or drink to balance this.

Your body temperature should be around 37°C.

Quick Test

1. Name three types of neurone.
2. Name four internal conditions that are kept relatively constant.
3. Name the hormone that causes insoluble glycogen to change to glucose.
4. What is the function of the thermoregulatory centre in the brain?

Hot Conditions:

If your core body temperature becomes too **high**…
- blood vessels in skin **dilate** (become wide), increasing heat loss
- sweat glands release **sweat** that evaporates, causing cooling.

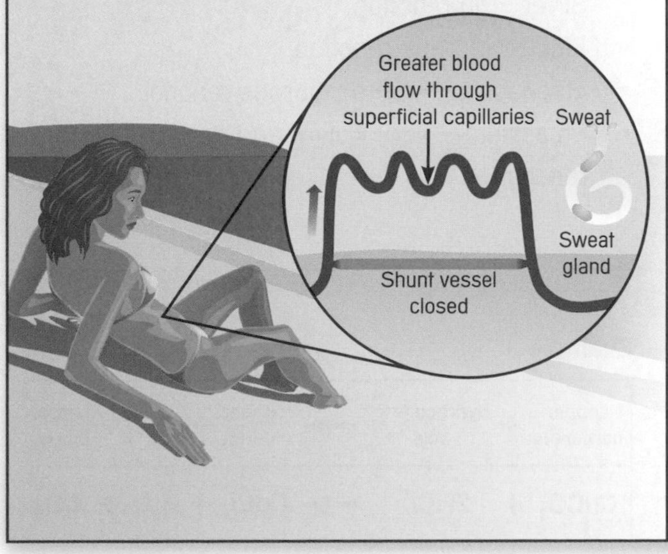

Greater blood flow through superficial capillaries   Sweat

Shunt vessel closed

Sweat gland

Cold Conditions:

If your core body temperature becomes too **low**…
- blood vessels in skin **constrict** (become narrower), reducing heat loss
- muscles start to '**shiver**' causing heat energy to be released by respiration in cells.

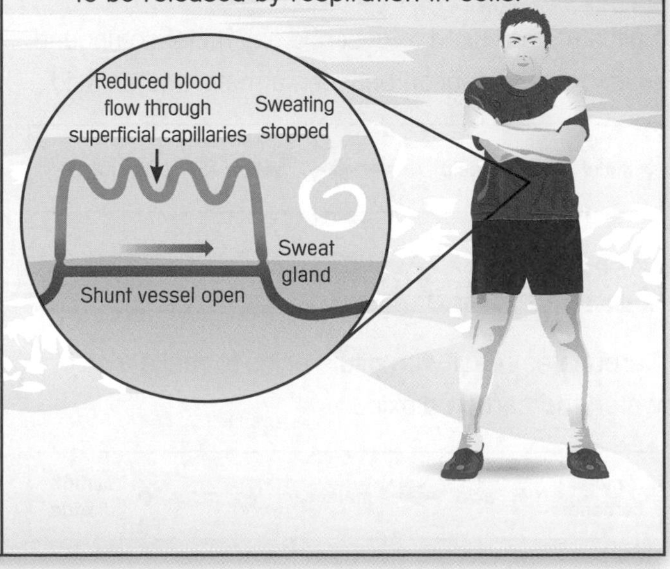

Reduced blood flow through superficial capillaries   Sweating stopped

Shunt vessel open

Sweat gland

U2 Chemistry in Action in the Body

Chemical Reactions in the Body

The body functions properly due to a series of complex **chemical reactions**.

Acids and Alkalis

Acids and **alkalis** are chemical opposites:
- Acids contain hydrogen **ions**, $H^+(aq)$.
- Alkalis contain hydroxide ions, $OH^-(aq)$.

Some acids and bases can be irritants. If they come into contact with your skin, you must wash them off in order to prevent irritation.

Concentrated acids and alkalis are corrosive and will damage skin or metal.

Harmful chemicals have warning symbols on their bottles.

Symbol	Name	Symbol	Name
	Toxic		Corrosive
	Oxidising		Irritant
	Harmful		Radioactive
	Highly flammable		Harmful to the environment

Stomach Acid

Our stomach contains dilute hydrochloric acid, which helps enzymes to digest food. Stomach enzymes work most effectively in an acidic pH.

Antacids, which are bases, are used to treat heart burn and nausea (sickness) caused by too much stomach acid. These antacids neutralise excess stomach acid.

Soluble Salts from Solids

Metals react with dilute acid to form a **metal salt** and **hydrogen**.

Salt is a word used to describe any metal compound made from a reaction between a metal and an acid.

metal	+	acid	⟶	salt	+	hydrogen

Some metals react with acid more vigorously than others:
- Silver – no reaction.
- Zinc – vigorous reaction.
- Magnesium – more vigorous reaction.
- Potassium – very violent and dangerous reaction.

Soluble Salts from Carbonates

Carbonates react with dilute acids to make a salt, water and carbon dioxide.

metal carbonate	+	acid	⟶	metal salt	+	water	+	carbon dioxide

Example

copper carbonate	+	hydrochloric acid	⟶	copper chloride	+	water	+	carbon dioxide
$CuCO_3$	+	$2HCl$	⟶	$CuCl_2$	+	H_2O	+	CO_2

Soluble Salts from Insoluble Bases

Bases are the oxides and hydroxides of metals. Soluble bases are called **alkalis**.

The oxides and hydroxides of many metals are **insoluble**. Their salts are prepared in the following way:

1 The metal oxide or hydroxide is added to an acid until no more will react.

2 The excess metal oxide or hydroxide is then filtered, leaving a solution of the salt.

3 The salt solution is then evaporated.

This reaction can be written generally as follows:

Example

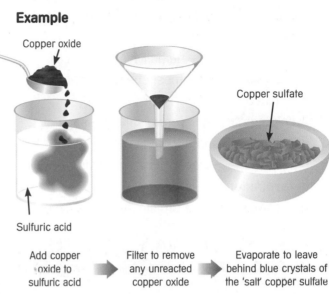

Copper oxide

Copper sulfate

Sulfuric acid

| Add copper oxide to sulfuric acid | Filter to remove any unreacted copper oxide | Evaporate to leave behind blue crystals of the 'salt' copper sulfate |

Salts of Alkali Metals

Compounds of alkali metals, called **salts**, can be made by reacting solutions of their hydroxides (which are alkaline) with a particular acid. This neutralisation reaction can be represented as follows:

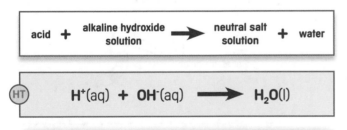

HT $H^+(aq) + OH^-(aq) \longrightarrow H_2O(l)$

The salt produced depends on the metal in the alkali and the acid used.

	Hydrochloric Acid	Sulfuric Acid	Nitric Acid
Sodium Hydroxide	Sodium chloride and water	Sodium sulfate and water	Sodium nitrate and water
Potassium Hydroxide	Potassium chloride and water	Potassium sulfate and water	Potassium nitrate and water

Insoluble Salts

Insoluble salts can be made by mixing appropriate solutions of ions so that a precipitate (solid substance) is formed.

Precipitation can be used to remove unwanted ions from a solution, e.g. softening hard water. The calcium (or magnesium) ions are precipitated out as insoluble calcium (or magnesium) carbonate.

Quick Test

1 Name the ions that react together in a neutralisation reaction.

2 What type of chemical is an antacid?

3 What is the name of the gas produced when a metal reacts with an acid?

4 What type of acid would you use if you wanted to make a metal sulfate?

U2 Human Inheritance and Genetic Disorders

Animal Cells

Human cells, most animal cells and most plant cells have the following parts:

- **Nucleus** – controls the activities of the cell.
- **Cytoplasm** – where most chemical reactions take place.
- A **cell membrane** – controls the passage of substances in and out of the cell.
- **Mitochondria** – where most energy is released in **respiration**.
- **Ribosomes** – where protein synthesis occurs.

A Human Cheek Cell

Cell membrane
Mitochondria
Cytoplasm (contains mitochondria)
Nucleus
Ribosomes

Genetic Information

The nucleus of a cell contains **chromosomes**.

- **Chromosomes** are made up of a substance called **DNA**.
- A section of a chromosome is called a **gene**.
- Different **genes** control the development of different **characteristics**.

During **reproduction**, genes are passed from parent to offspring (they are **inherited**).

Chromosomes come in **pairs**, but different species have different numbers of pairs. Humans have **23 pairs**.

A Section of One Chromosome

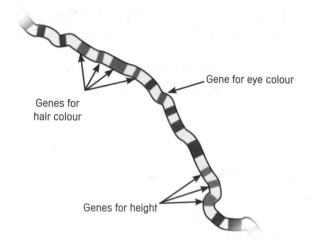

Genes for hair colour
Gene for eye colour
Genes for height

Variation

Differences between individuals of the same species are called **variation**. Variation may be due to **genetic** or **environmental factors**.

For example:

- **Genetic** factors are responsible for the colour of dogs' coats being different.
- **Environmental** factors are responsible for identical twins being very different in weight. If one twin has a diet high in fat and doesn't exercise, he will become fatter than his brother.

Key Words Nucleus • Mitochondria • Respiration • Ribosome • Chromosome • DNA • Gene • Variation

Human Body Cells

Body cells contain **46 chromosomes** arranged as **23 pairs**.

Gametes are sex cells, i.e. female eggs and male sperm. Gametes have **23 chromosomes** (one from each pair).

The fusion of two gamete cells produces a **zygote**. A zygote has 46 chromosomes in total (23 pairs).

Pairs of Chromosomes in a Male

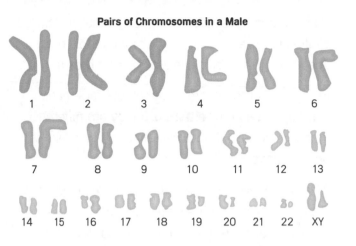

Inheritance of the Sex Chromosome

Of the 23 pairs of chromosomes in the human body, 1 pair is called the **sex chromosomes**.

- **In females**, these chromosomes are identical and are called the X chromosomes.
- **In males**, one chromosome (Y) is much shorter than the other (X).

Female Sex Chromosomes	Male Sex Chromosomes
X X	X Y

Like all pairs of chromosomes, offspring inherit...
- one sex chromosome from the mother
- one sex chromosome from the father.

The sex of an individual is ultimately decided by whether the ovum is fertilised by an **X-carrying sperm** or a **Y-carrying sperm**.

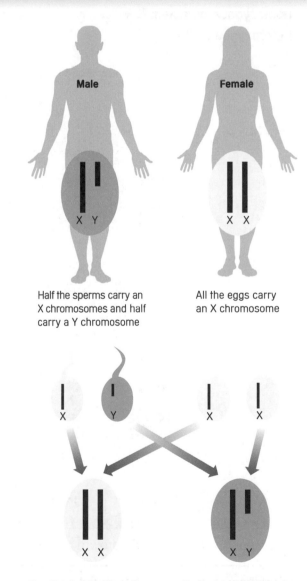

Half the sperms carry an X chromosomes and half carry a Y chromosome

All the eggs carry an X chromosome

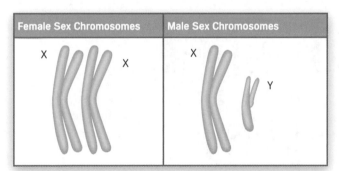

If an X sperm fertilises the egg, a girl is produced

If a Y sperm fertilises the egg, a boy is produced

Alleles

Some **genes** have different forms or variations, called **alleles**. For example, the gene that controls one type of eye colour has two alleles – blue or brown.

In a pair of chromosomes, the alleles for a gene can be the **same** or **different**. If they are different, then one allele will be **dominant** and one allele will be **recessive**.

A dominant allele **will control** the characteristics of the gene. Dominant alleles express themselves even if present on only one chromosome. In the example of brown eyes, an individual can be…
* **homozygous dominant** (BB)
* **heterozygous** (Bb).

A recessive allele will **only** control the characteristics of the gene if it is present on **both chromosomes in a pair** (i.e. no dominant allele is present). So, in the example of blue eyes, an individual can only be **homozygous recessive** (bb).

Example: Dominant and Recessive Alleles
The diagram shows two pairs of genes from the middle of a pair of chromosomes. The genes code for…
* eye colour
* type of earlobe (i.e. attached or unattached).

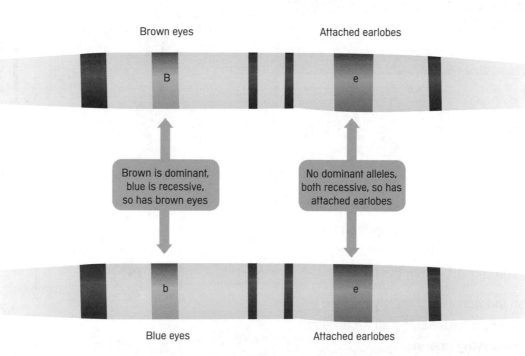

Dominant and Recessive Genes

Capital letter for dominant allele
Lower case letter for recessive allele

Brown eyes

Attached earlobes

B

e

Brown is dominant, blue is recessive, so has brown eyes

No dominant alleles, both recessive, so has attached earlobes

b

e

Blue eyes

Attached earlobes

	Homozygous Dominant	Heterozygous	Homozygous Recessive
Eye colour	BB (brown)	Bb (brown)	bb (blue)
Earlobes	EE (free earlobes)	Ee (free earlobes)	ee (attached earlobes)

Monohybrid Inheritance

When a characteristic is determined by **just one pair of alleles** then a simple **genetic cross diagram** can be drawn to show inheritance of genes.

This type of inheritance is referred to as **monohybrid inheritance**.

Inheritance of Eye Colour

In genetic diagrams you should use **capital letters for dominant alleles** and **lower case letters for recessive alleles**. So, for eye colour, 'B' is used for brown eye alleles and 'b' is used for blue eye alleles.

From the crosses on the diagrams, the following can be seen:

❶ If one parent has two dominant alleles then **all the offspring will** inherit that characteristic.

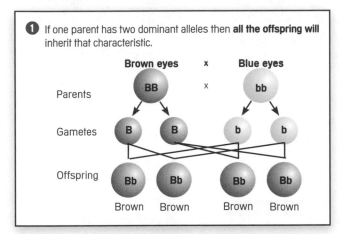

❸ If one parent has one recessive allele and the other has two recessive alleles, then there is a 50% chance of that characteristic appearing in the offspring.

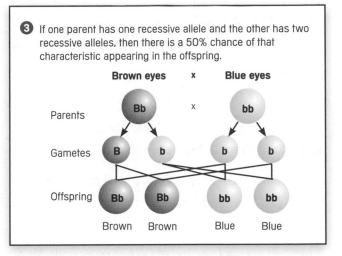

But remember, these are **only probabilities**. In practice, all that matters is which egg is fertilised by which sperm. This process is completely **random**.

This information can also be represented using a Punnett square.

	Gametes formed by brown-eyed parent	
Gametes formed by blue-eyed parent	B	b
b	Bb	bb
b	Bb	bb

When you construct genetic diagrams remember to...
- clearly **identify** the alleles of the parents
- place each of these alleles in a **separate gamete**
- **join** each gamete with the gametes from the other parent.

❷ If both parents have one recessive allele then this characteristic **may** appear in the offspring (25% chance).

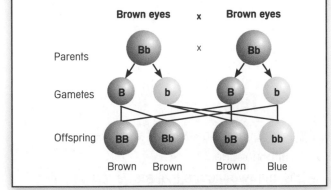

Differentiation of Cells

Differentiation is when cells develop a **specialised** structure to carry out a specific function:

- Most plant cells can differentiate throughout life.
- Animal cells **can only** differentiate at an early stage.
- Mature animal cells divide for repair / replacement.

U2 Human Inheritance and Genetic Disorders

Stem Cells

Stem cells are cells in human **embryos** and adult **bone marrow** that have **not yet differentiated**.

They can differentiate into **many different types** of cells, e.g. nerve cells. Treatment using these cells may help treat people with conditions like **paralysis**.

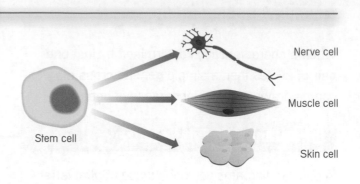

Nerve cell

Muscle cell

Stem cell

Skin cell

Chromosomes, DNA and Genes

Chromosomes are made up of long, coiled molecules of **DNA (deoxyribonucleic acid)**.

A DNA molecule consists of **two strands** that are coiled to form a **double helix**.

Each person has **unique** DNA (apart from identical twins). So, DNA can be used for **identification** (DNA fingerprinting).

A **gene** is a small section of DNA. Genes code for a particular inherited characteristic, e.g. eye colour.

Genes code for a particular characteristic by providing a code for a combination of amino acids that make up a protein.

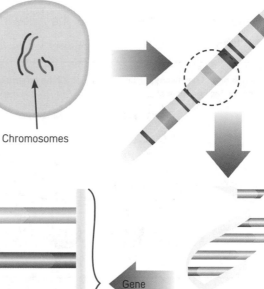

A Cell

A Section of Chromosome

Chromosomes

Gene

A Section of Uncoiled DNA (part of a gene)

A Section of DNA

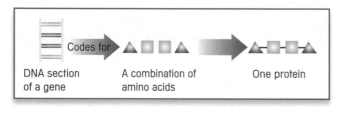

Codes for

DNA section of a gene

A combination of amino acids

One protein

Genetic Disorders

Embryos can be screened for the genes that cause genetic disorders. For example:

- **Polydactyly** is a disorder where a person may have more than five fingers or toes on each hand / foot.
- **Cystic fibrosis** is a disorder of cell membranes.
- **Haemophilia** is a disorder of the blood.
- **Sickle-cell anaemia** is a disorder of red blood cells.

Quick Test

1. Name the process that takes place in the mitochondria.
2. Name two factors that may affect variation.
3. How many chromosomes do human body cells contain?
4. What is an allele?

Limestone

Limestone (CaCO$_3$) is a **sedimentary rock** that is obtained from the ground by quarrying:

- It consists mainly of **calcium carbonate**.
- It is easy to obtain and has many uses.

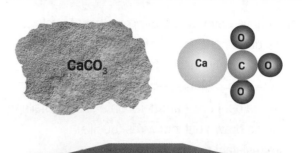

Limestone can be used...

- as a building material
- for producing slaked lime
- for making glass
- for making cement, mortar and concrete.

As a Building Material

Limestone can be quarried, cut into blocks and used to build houses. It can be **eroded** by **acid rain**, but this is a very slow process.

Producing Slaked Lime

When calcium carbonate (CaCO$_3$) is heated in a kiln it decomposes. This reaction is called **thermal decomposition**. It causes the calcium carbonate to break down into calcium oxide (quicklime, CaO) and carbon dioxide. The calcium oxide can then be reacted (**slaked**) with water to produce **slaked lime** (calcium hydroxide, Ca(OH)$_2$). Slaked lime can be used to neutralise lakes and acid soils, preventing crop failure.

Carbonates of other metals decompose in a similar way when they are heated.

Making Glass

Powdered limestone is mixed with sand and sodium carbonate and heated. When it cools, it is transparent.

Making Cement, Mortar and Concrete

Powdered limestone is roasted in a rotary kiln with powdered clay to produce dry **cement**. When sand and water are mixed in, **mortar** is produced. Mortar is used to hold bricks and stones together. When gravel, sand and water are mixed in, **concrete** is produced.

U2 Materials Used to Construct Our Homes

Metals

Metals have several useful properties. Metals…

- are good conductors of heat and electricity
- are relatively hard and mechanically strong
- have high melting points (except mercury)
- can be bent and hammered into shape (malleable)
- can be drawn out into wires (ductile)
- are resistant to corrosion.

The layers of atoms in metals are able to slide over each other. This means that metals can be **bent and shaped**.

Delocalised electrons in and around the metal can move freely through the metal. This makes them good conductors of electricity.

Using Metals

Metals have many uses.

Name of Metal	Use	Property
Copper	• Water pipes • Hot water cylinders • Wiring	• Malleable, strong, high melting point • Good conductor of electricity
Lead	• Flashing on roofs	• Unreactive and malleable
Steel	• Supporting structures and fixings	• High tensile strength
Aluminium	• Window frames	• Resistant to corrosion • Malleable • Strong and light

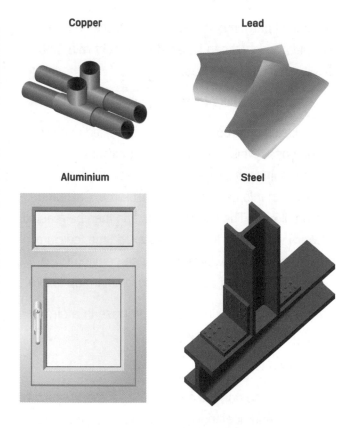

Copper

Lead

Aluminium

Steel

Using Ceramics

Ceramics are made by heating inorganic compounds, e.g. silica, alumina and clay. Inorganic compounds are compounds that don't contain carbon.

Ceramics are **giant structures** of metallic and non-metallic elements that are held together by either **ionic** or **covalent bonds**, depending on the atoms present.

In a giant structure (lattice), every atom is bonded to several others around it. For example, china and pottery.

Glass, bricks and cement have many uses, for example…

- in buildings
- as components of **composite** materials.

Key Words   **Ionic bond • Covalent bond • Composite**

Properties of Ceramics

Ceramics…
- are **hard** and **strong** because the lattice is held together by many bonds
- are **brittle** because the atoms can't slide past each other, which means an applied force breaks the lattice apart
- have a **high melting point** because a lot of energy is needed to break the strong bonds that hold the lattice together
- have **poor thermal conductivity** and **poor electrical conductivity** because the electrons are fixed in bonds and aren't free to move.

Composites

Composites are made by combining **two or more** materials to make use of the best properties of each material. For example, they can be formed in **layers** or by **embedding** one material in the 'matrix' of another.

Examples:
- **GRP (glass fibre-reinforced polyester) or fibreglass** is made of polyester resin with glass fibres embedded in it to strengthen it. It is used to make tennis racquets, surfboards and skis.
- **Plasticised polyvinylchloride (PVC)** is used for the soles of trainers. The stiff interlocking PVC molecules have plasticiser molecules between them, so the material becomes **flexible**.
- **Carbon fibre-reinforced epoxy resin** is used for mountain bikes and tennis racquets. Carbon fibres are embedded in epoxy resin to strengthen it, whilst keeping it **lightweight**.

The Structure of a Composite

Polymer chain

Plasticiser molecule

Examples of Composites

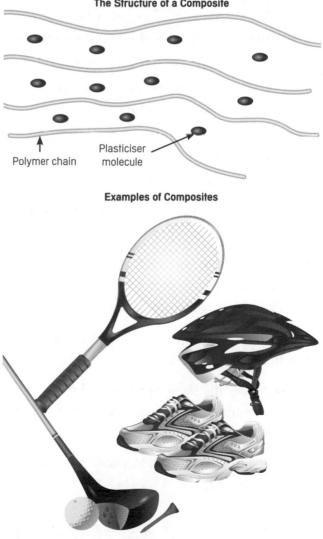

Properties of Composites

Composites are made from a combination of two or more materials, so their **properties** depend on the properties of the materials they were made from.

Alkenes (Unsaturated Hydrocarbons)

Carbon atoms can form **double bonds**. This means that not all the carbon atoms are linked to four other atoms; a **double carbon–carbon bond** is present instead.

Long hydrocarbon chain molecules can be split or 'cracked' to make several smaller shorter hydrocarbon chain molecules.

Some of the products of cracking are hydrocarbon molecules with at least one double bond (alkenes):
- The general formula for alkenes is C_nH_{2n}.
- The simplest alkene is ethene, C_2H_4.
- Ethene is made up of four hydrogen atoms and two carbon atoms, and contains one double carbon–carbon bond.

Alkenes can be represented using displayed formulae:

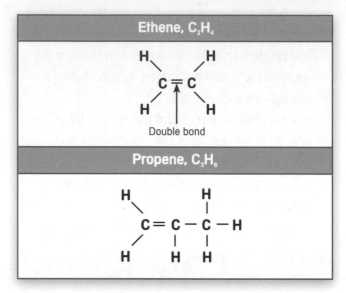

Ethene, C_2H_4

Double bond

Propene, C_3H_6

Polymerisation

Because alkenes are unsaturated (they have a double bond), they are useful for making other molecules, especially **polymers** (long-chain molecules).

Small alkene molecules (monomers) join together to form polymers. This is **polymerisation**.

Polymers such as **poly(ethene)** and **poly(propene)** are made in this way.

The properties of polymers depend on…
- what they are made from
- the conditions under which they are made.

For example, slime with different **viscosities** can be made from poly(ethenol).

The viscosity of the slime depends on the temperature and concentrations of the poly(ethenol) and borax from which it is made.

The materials commonly called **plastics** are all synthetic polymers.

The small alkene molecules are called monomers

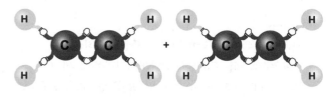

Their double bonds are easily broken

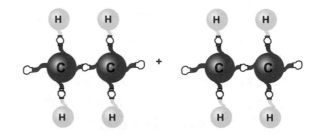

Large numbers of molecules can therefore be joined in this way

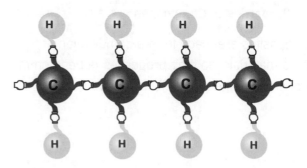

Polymer • Polymerisation

Representing Polymerisation

Polymerisation can be represented like this:

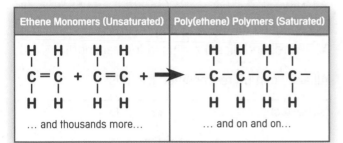

Ethene Monomers (Unsaturated)	Poly(ethene) Polymers (Saturated)
... and thousands more...	... and on and on...

The general formula for polymerisation can be used to represent the formation of any simple polymer:

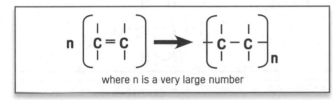

where n is a very large number

Polymers

Polymers...

- are flexible
- are poor conductors of heat and electricity
- are resistant to corrosion
- are waterproof
- are materials with low melting points.

Polymers have many useful applications and new uses are being developed.

Polymers and composites are widely used in medicine and dentistry. For example...

- implantable materials are used for tissue surgery
- hard-wearing anti-bacterial dental cements, coating and fillers are used in dentistry
- hydrogels can be used as wound dressings.

Polymers can be used to coat fabrics with a waterproof layer.

Smart materials, including shape-memory polymers, are increasingly more common.

Specific polymers can have different uses. For example...

- PVC is used to make waterproof items, drain pipes, and can be used as electrical insulators
- polystyrene is used to make the casing for electrical appliances, and it can be expanded to make protective packaging
- poly(ethene) is used to make plastic bags and bottles
- poly(propene) is used to make crates and ropes.

Quick Test

1. Which property of copper metal makes it useful for electrical wires?
2. How are ceramics made?
3. Name three examples of composites.
4. Alkenes are unsaturated. What does unsaturated mean?

Alkanes (Saturated Hydrocarbons)

The 'spine' of a hydrocarbon is made up of a chain of carbon atoms. When these are joined together by **single carbon–carbon bonds**, the hydrocarbon is **saturated** and is known as an **alkane**.

- Hydrogen atoms can make one bond each.
- Carbon atoms can make four bonds each.
- The simplest alkane, **methane**, is made up of four hydrogen atoms and one carbon atom.

The general formula for alkanes is C_nH_{2n+2}

The carbon atoms in alkanes are linked to four other atoms by **single bonds**. This means that the alkane is saturated. This explains why alkanes are fairly unreactive, but they burn well.

The shorter-chain hydrocarbons release energy more quickly by burning, because they are more volatile, so there is greater demand for them as **fuels**.

Fuels that we use include the following:
- Natural gas, kerosene (paraffin) and heating oil are used for cooking and heating.
- Petrol and diesel are used for transport.

Alkanes can be represented like this:

Methane, CH_4	
	H $\|$ $H - C - H$ $\|$ H

Ethane, C_2H_6	
	$H \quad H$ $\| \quad \|$ $H - C - C - H$ $\| \quad \|$ $H \quad H$

Propane, C_3H_8	
	$H \quad H \quad H$ $\| \quad \| \quad \|$ $H - C - C - C - H$ $\| \quad \| \quad \|$ $H \quad H \quad H$

Burning Fuels

Most **fuels** contain carbon and hydrogen. Many also contain **sulfur**. As fuels burn they produce waste products, which are released into the atmosphere:

carbon	burn with oxygen →	carbon dioxide
C + O_2	→	CO_2

carbon	burn with less oxygen →	carbon monoxide
$2C$ + O_2	→	$2CO$

hydrogen	burn with oxygen →	water vapour
$2H_2$ + O_2	→	$2H_2O$

sulfur	burn with oxygen →	sulfur dioxide
S + O_2	→	SO_2

(HT) methane + oxygen → carbon dioxide + water + heat energy

$$CH_4 + 2O_2 \rightarrow CO_2 + 2H_2O + \text{heat energy}$$

For example, the fuel methane (CH_4) burns to produce carbon dioxide (from the carbon in the methane) and water (from the hydrogen in the methane).
- Carbon dioxide causes **global warming** due to the greenhouse effect.
- Sulfur dioxide causes **acid rain**.

Sulfur can be removed from fuel before burning (e.g. in motor vehicles). Sulfur dioxide can be removed from the waste gases after **combustion** (e.g. in power stations). Both of these processes add to the cost.

Quick Test

1 Name the two elements found in hydrocarbons.
2 Alkanes are described as saturated hydrocarbons. What does saturated mean?
3 Name the two chemicals produced when methane completely burns in oxygen.

Fuels

Fuels are substances that release useful amounts of **energy** when they are burned.

Coal   Oil   Gas

Non-Renewable Energy Sources

Coal, oil and **natural gas** are **fossil fuels**. They are **energy sources** that we depend on for most of the energy that we use.

They can't be replaced within a lifetime, so they will eventually run out. They are **non-renewable energy sources**.

Nuclear fuels such as **uranium** and **plutonium** are also **non-renewable**.

Nuclear fission is the splitting of an unstable nucleus into two or more smaller nuclei. This releases neutrons that can collide with nuclei, splitting them. This can cause a chain reaction that generates huge amounts of heat energy.

Nuclear fuel isn't burned like coal, oil or gas to release energy and it isn't classed as a fossil fuel. Non-renewable energy sources can be used to generate electricity.

Fossil fuels (and wood) are used to generate electricity in power stations:

1. Fossil fuels (and wood) are burned to release **thermal energy**.
2. The thermal energy boils water to produce steam.
3. The steam drives turbines that are attached to an electrical generator.

N.B. Wood isn't a fossil fuel and is renewable.

Nuclear fuel is used to generate electricity in a similar way:

1. A reactor is used to generate heat by nuclear fission.
2. A heat exchanger transfers the thermal energy from the reactor to the water.
3. The water turns to steam and drives the turbines.

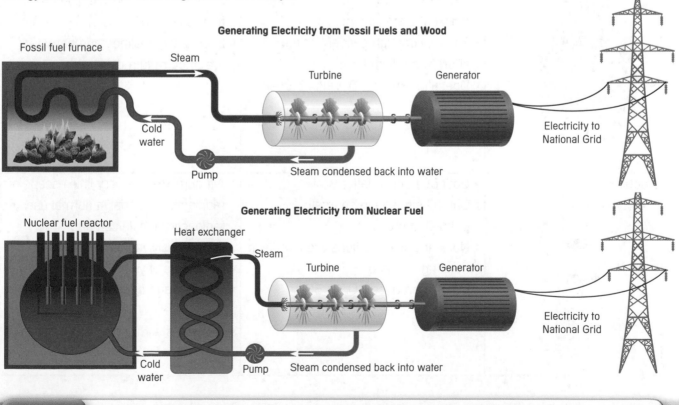

Generating Electricity from Fossil Fuels and Wood

Fossil fuel furnace — Steam — Turbine — Generator — Electricity to National Grid — Cold water — Pump — Steam condensed back into water

Generating Electricity from Nuclear Fuel

Nuclear fuel reactor — Heat exchanger — Steam — Turbine — Generator — Electricity to National Grid — Cold water — Pump — Steam condensed back into water

Non-renewable Energy Sources

Source	Advantages	Disadvantages
Coal	• Relatively cheap and easy to obtain. • Coal-fired power stations have a relatively quick start-up time. • There may be over a century's worth of coal left.	• Burning produces CO_2 and SO_2. • Produces more CO_2 per unit of energy than oil or gas does. • SO_2 causes **acid rain**. (Removing the SO_2 is costly.)
Oil	• Relatively easy to find, though the price is variable. • Oil-fired power stations are flexible in meeting demand. • Oil-fired power stations have a quick start-up time. • There is enough oil left for the short to medium term.	• Burning produces CO_2 and SO_2. • Produces more CO_2 per unit of energy than gas does. • Tankers pose the risk of **spillage** and **pollution**.
Natural gas	• Can be found as easily as oil. • Gas-fired power stations have the quickest start-up time. • There is enough gas left for the short to medium term. • Doesn't produce SO_2 when burned.	• Burning produces CO_2 (although it produces less per unit of energy than coal or oil). • Expensive pipelines and networks are often required to transport it.
Nuclear	• Cost of fuel is relatively low. • Can be situated in sparsely populated areas. • Nuclear power stations are flexible in meeting demand. • Doesn't produce CO_2 or SO_2.	• Although there is very little escape of radioactive material in normal use, radioactive waste can stay **dangerously radioactive** for thousands of years. • Building and decommissioning is costly. • Longest start-up time.

Renewable Energy Sources

Renewable energy sources will not run out; they are continually being replaced.

Many renewable energy sources are 'powered' by the **Sun** or **Moon**. The gravitational pull of the Moon creates tides.

The Sun causes…
- evaporation, which results in rain and flowing water
- convection currents, which result in winds, which create waves.

Renewable energy sources can be used to drive turbines or generators directly.

Wind Turbines Propeller blades Wind Generator	• Wind drives huge turbines, which drive generators. • Usually positioned on hills so they are exposed to as much wind as possible.
Solar Cells and Panels Solar energy Electricity out	• Made of a semiconductor material (usually silicon). • Absorb radiation from the Sun and transform it into electrical energy.
Hydro-Electric Dam Dam Reservoir Turbines	• Water stored in a reservoir above the power station flows down through pipes to drive turbines, which generate electricity. Rivers can also be used to drive turbines.
Tidal Barrage Barrage Water at high tide Turbines	• As the tide comes in, water flows through a valve in the barrage and becomes trapped. • At low tide, the water is released through a gap that has a turbine in it, which drives a generator.
Nodding Duck Rotating nodding duck	• Positioned in the sea. Wave motion makes the 'ducks' rock. • This movement is translated into a rotary movement, which drives a generator.

Geothermal
 Generating station
 Cold water
 Hot water and steam
	• In some volcanic areas…
	   – hot water and steam rise naturally to the surface
	   – cold water is pumped into hot rock to generate steam.
	 • Steam is used to drive turbines, which drive generators.

Biomass	• Biomass is plant or animal material that can be used as a fuel for heating and cooking. • Household waste, composed of waste material, can be used to heat swimming pools and homes.

Renewable Energy Sources (Cont.)

The energy sources listed below can provide clean, safe energy. Some of their advantages and disadvantages are given in the table.

Source	Advantages	Disadvantages
Wind	• No fuel and little maintenance required. • No pollutant gases produced. • Can be built offshore.	• Cause noise and visual pollution. • Not very flexible in meeting demand (unless energy is stored). • Initial high capital outlay to build them. • Dependent on the wind
Tidal and Waves	• No fuel required. • No pollutant gases produced. • Barrage water can be released when electricity demand is high.	• They are unsightly, a hazard to shipping, and destroy habitats. • Variations of tides and waves affect output. • Initial high capital outlay to build them.
Hydro-Electric	• Fast start-up time. • No pollutant gases produced. • Water can be pumped back up to the reservoir when electricity demand is low.	• Location is critical: often involves damming upland valleys. • There must be adequate rainfall in the region where the reservoir is. • Very high initial capital outlay.
Solar	• Can produce electricity in remote locations. • No pollutant gases produced.	• Dependent on intensity of light. • High cost per unit of electricity produced.

Summary

Advantages of Using Renewable Energy Resources	Disadvantages of Using Renewable Energy Resources
• No fuel costs during operation. • Very little chemical pollution. • Often low maintenance.	• Produce small amounts of electricity (except hydro-electric). • Take up lots of space and are unsightly. • Unreliable (apart from hydro-electric) – they depend on the weather. • High initial capital outlay.

The National Grid

Electricity is generated at power stations. It is then transferred to homes, schools and factories by a network of cables called the **National Grid**.

Transformers are used to change the **voltage** of the **alternating current** supply before and after it is transmitted through the National Grid.

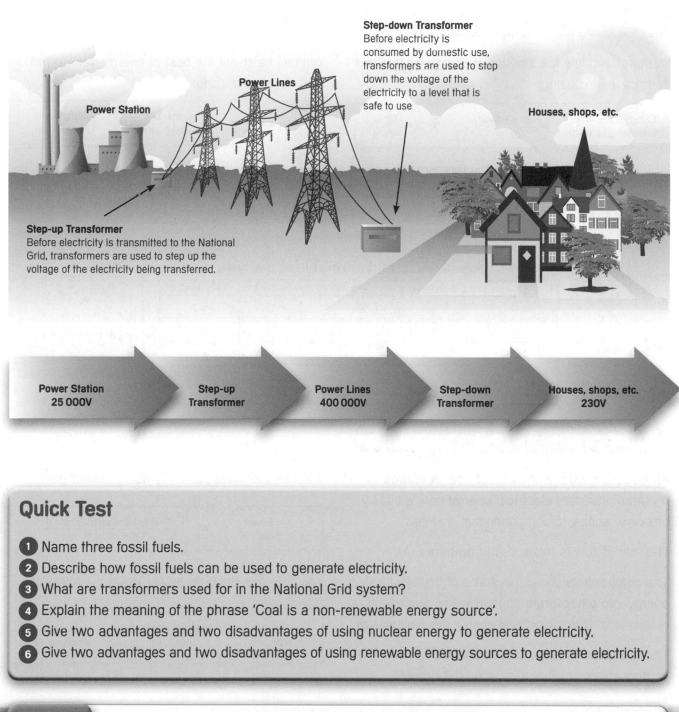

Step-down Transformer
Before electricity is consumed by domestic use, transformers are used to step down the voltage of the electricity to a level that is safe to use

Power Lines

Power Station

Houses, shops, etc.

Step-up Transformer
Before electricity is transmitted to the National Grid, transformers are used to step up the voltage of the electricity being transferred.

Power Station 25 000V	→	Step-up Transformer	→	Power Lines 400 000V	→	Step-down Transformer	→	Houses, shops, etc. 230V

Quick Test

1. Name three fossil fuels.
2. Describe how fossil fuels can be used to generate electricity.
3. What are transformers used for in the National Grid system?
4. Explain the meaning of the phrase 'Coal is a non-renewable energy source'.
5. Give two advantages and two disadvantages of using nuclear energy to generate electricity.
6. Give two advantages and two disadvantages of using renewable energy sources to generate electricity.

Key Words   **National Grid • Transformer • Voltage**

U2 The Cost of Running Appliances in the Home

Energy Transformation

Most of the **energy** transferred to homes and industry is **electrical energy**.

Electrical energy is easily transformed into…
- **heat / thermal energy** (e.g. hairdryer)
- **light energy** (e.g. lamp)
- **sound energy** (e.g. stereo speakers)
- **kinetic / movement energy** (e.g. electric fan).

The amount of energy transformed by an electrical appliance depends on…
- how long the appliance is switched on
- how fast the appliance can transform energy.

The **power** of an appliance is measured in **watts** (W) or **kilowatts** (kW).

Energy is normally measured in **joules** (J).

Energy Calculations

You can calculate the **amount of energy transferred** from the mains using:

Energy transferred (kilowatt-hour, kWh)	=	Power (kilowatt, kW)	X	Time (hour, h)

You can calculate the **cost of energy transferred** from the mains using:

Total cost	=	Number of kilowatt-hours	X	Cost per kilowatt-hour

Normal reading – electricity used

kWh calculated by subtracting present reading from previous reading

REB Regional Electricity Board

Mr R. Jones
273 Dove Street
Southampton
SW15 WFK

Electricity Statement. Period: 01.01.11 – 31.01.11 No standing charge

Present reading	Previous reading	kWh used	Cost per kWh (p)	Charge amount (£)
13640	13890	250	18.5	46.25
			Total VAT exclusive charges	46.25
			VAT at 5%	2.31
			Total charges including VAT	48.56

Regional Electricity Board, Anchor House, Ingleby Street, Southampton SW15 TNE **Telephone:** 01445 680180 **Fax:** 01445 680180 **Email:** info@reb.co.uk **Web:** www.reb.co.uk

Kilowatt-hours is the unit of power

Cost
= kWh used x cost per kWh
= 250 x 18.5p

VAT = 46.25 x $\frac{5}{100}$

Power

An electric **current** is the rate of **flow of charge**. A current transfers **electrical energy** from a battery or power supply to components in a circuit.

The rate of flow is measured in **amperes** (A).

The **components transform** some of this electrical energy into **other forms** of energy. For example, a **resistor** transforms **electrical** energy into **heat energy**.

The rate at which energy is transformed by a component or appliance is called the **power**.

You can **calculate** power by using this formula:

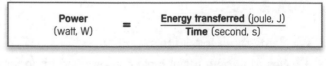

Power (watt, W)	=	Energy transferred (joule, J) / Time (second, s)

You can also calculate power using the formula:

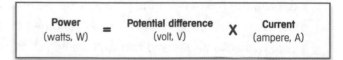

Power (watts, W)	=	Potential difference (volt, V)	X	Current (ampere, A)

Power • Current • Transform • Resistor

Transferring and Transforming Energy

When devices transfer energy, only part of the energy is usefully transferred to where it's wanted and in the form that's wanted.

The remaining energy is **transferred** in a non-useful way, mainly as heat energy. It is known as **wasted energy**.

For example, a light bulb transforms electrical energy into useful light energy. However, most of the energy is wasted as heat energy.

The wasted energy and the useful energy are eventually transferred to the surroundings, which become warmer.

No energy is created or destroyed. It is just changed into a different form (transformed).

Often, the energy becomes increasingly spread out, so it's difficult for any further useful energy transfers to occur.

Efficiency

The **efficiency** of a device refers to the proportion of energy that is usefully transformed. The greater the proportion of energy that is usefully transformed, the more **efficient** the device is.

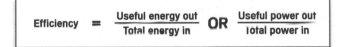

$$\text{Efficiency} = \frac{\text{Useful energy out}}{\text{Total energy in}} \quad \textbf{OR} \quad \frac{\text{Useful power out}}{\text{Total power in}}$$

For example, only a quarter of the energy supplied to a television is usefully transformed into light and sound. So it's only 25% efficient.

Energy labels on appliances provide information about the efficiency of energy conversions. Appliances with high efficiency are more economical to run but can be more expensive to buy.

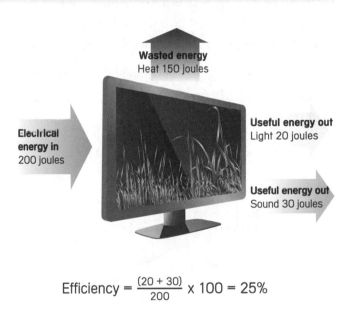

Wasted energy
Heat 150 joules

Electrical energy in
200 joules

Useful energy out
Light 20 joules

Useful energy out
Sound 30 joules

$$\text{Efficiency} = \frac{(20 + 30)}{200} \times 100 = 25\%$$

Sankey Diagrams

Sankey diagrams can be used to show the type of energy transformations and transfers that take place through an electrical appliance, e.g. a television. They are drawn to scale using a grid.

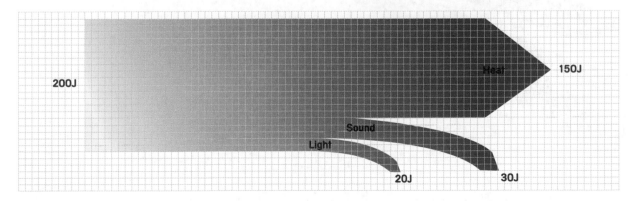

200J

Heat   150J

Sound

Light

20J   30J

U2 Electromagnetic Waves in the Home

Electromagnetic Radiation

Electromagnetic radiations are disturbances in an **electric field**.

They travel as **waves** and move **energy** from one place to another. Each type of electromagnetic radiation...
- has a different **wavelength**
- has a different **frequency**.

All types of electromagnetic radiation travel at the same speed through a **vacuum**. Electromagnetic radiations form the **electromagnetic spectrum**.

Different wavelengths of electromagnetic radiation are **reflected, absorbed** or **transmitted** in different ways by different substances and types of surface.

When a wave (radiation) is absorbed by a substance the energy is absorbed and makes the substance heat up.

Electromagnetic waves obey the wave formula:

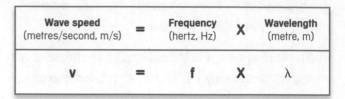

Wave speed (metres/second, m/s)	=	Frequency (hertz, Hz)	X	Wavelength (metre, m)
v	=	f	X	λ

Visible light is one type of electromagnetic radiation. The seven 'colours of the rainbow' form the **visible spectrum** (the only part of the electromagnetic spectrum that can be seen by humans).

The visible spectrum is produced because white light is made up of different colours. The colours are **refracted** by different amounts as they pass through a prism:
- **Red light** is refracted **the least**.
- **Violet light** is refracted **the most**.

The Electromagnetic Spectrum

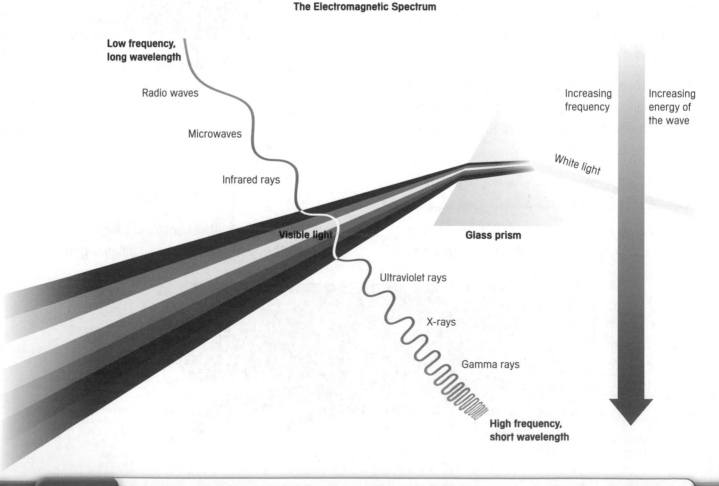

Low frequency, long wavelength

Radio waves

Microwaves

Infrared rays

Visible light

Ultraviolet rays

X-rays

Gamma rays

High frequency, short wavelength

Increasing frequency

Increasing energy of the wave

White light

Glass prism

Key Words   Radiation • Electromagnetic spectrum • Reflection • Absorption • Transmission • Refraction

Electromagnetic Waves	Uses	Effects
Radio Waves	• Transmitting radio and TV signals between places across the Earth.	• High levels of exposure for short periods can increase body temperature, leading to tissue damage.
Microwaves	• Satellite communication networks and mobile phone networks (they can pass through the Earth's atmosphere). • Cooking – water molecules absorb microwaves and heat up food.	• May damage or kill cells because microwaves are absorbed by water in the cells, which heat up.
Infrared Rays	• Grills, toasters and radiant heaters (e.g. electric fires). • Remote controls for televisions and dvd players.	• Absorbed by skin and felt as heat. • An excessive amount can cause burns.
Visible light	• **Optical fibre** communication.	
Ultraviolet Rays	• Security coding – special paint absorbs UV and emits visible light. • Sun tanning and sunbeds.	• Passes through skin to the tissues below. • High doses can kill cells. • Low doses can cause **cancer**.
X-Rays	• Producing shadow pictures of bones and metals. • Treating certain cancers.	• Passes through soft tissues (some is absorbed). • High doses can kill cells. • Low doses can cause cancer.
Gamma Rays	• Killing cancerous cells. • Killing bacteria on food and surgical instruments.	• Passes through soft tissues (some is absorbed). • High doses can kill cells. • Low doses can cause cancer.

Quick Test

1. Which equation could you use to work out the efficiency of a device?
2. What does the amount of energy transformed by a device depend on?
3. Give the equation that could be used to work out the total cost of energy transferred.
4. Which type of electromagnetic radiation has the lowest frequency and the longest wavelength?

U2 Exam Practice Questions

1 **a)** Our nervous system detects and responds to our surroundings.
Match the words A, B, C and D with the parts labelled 1 to 4 in the diagram below.

A receptor ....

B effector ....

C stimulus ....

D spinal cord .... **(4 marks)**

b) What is the name of the tiny gap between neurones? .... **(1 mark)**

2 Some examples of chemical reactions are shown below. **(5 marks)**

a) Complete the following word equations.

i) hydrochloric acid + potassium hydroxide ⟶ .... + water

ii) copper oxide + ... ⟶ copper sulfate + water

iii) .... + nitric acid ⟶ sodium nitrate + water

iv) barium chloride + zinc sulfate ⟶ .... + ...

b) What is the name of the only insoluble salt in the equations above? .... **(1 mark)**

3 The diagram shows a human cheek cell.

a) Give the names of the parts labelled A to C.

A ....

B ....

C .... **(3 marks)**

b) Explain what happens in mitochondria.

.... **(1 mark)**

c) Which structure controls the passage of substances in and out of the cell? .... **(1 mark)**

4 **a)** What is meant by the term 'variation'?

.... **(1 mark)**

b) Describe two factors that can affect a characteristic.

.... **(2 marks)**

c) Eye colour (brown or blue eyes) is controlled by one gene with two alleles.

- **B** is the allele that causes eyes to be brown and is the dominant allele.
- **b** is the allele that causes eyes to be blue and is the recessive allele.

A brown-eyed mother and a blue-eyed father have two children. One child has brown eyes and the other child has blue eyes. Describe and explain how this is possible. You might want to draw a diagram to help explain your answer. **(4 marks)**

5 Metals have many different uses because of their properties. Match the metals A–C with the descriptions numbered 1–3 below. **(3 marks)**

A iron (Fe) ....   **B** copper (Cu) ....

C aluminium (Al) ....

1 It is used to make the alloy steel.   **2** It is a good conductor of heat and electricity.
3 It is strong and light and will not rust.

6 The alkane ethane, C_2H_6, is a hydrocarbon fuel.
   a) Draw a structural formula diagram to illustrate its carbon–hydrogen and carbon–carbon bonds. **(2 marks)**

   b) i) Write a word equation to represent burning pure ethane.

   .... **(2 marks)**

   ii) If sulfur was present in an impure form of this fuel, which additional product would be made?

   .... **(1 mark)**

HT   **iii)** Write a balanced symbol equation to represent the burning of pure ethane.

   .... **(1 mark)**

Drugs

Drugs are chemical substances that alter the way your body works. Drugs can be **beneficial**, but they can also **harm** your body.

- Some drugs can be obtained from **natural** substances (many have been known to indigenous people for years).
- Some drugs are **synthetic** (developed by scientists).

Developing New Drugs

When new **medical drugs** are developed, they need to be thoroughly **tested** in the laboratory to find out if they are **toxic** (poisonous). Then they are checked for **side effects** in trials on human volunteers.

The flow chart shows the stages in developing a new drug.

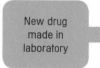

New drug made in laboratory	→	Tested in laboratory for toxicity	→	Trialled on volunteers to check for side effects	→	Passed for use by the Medicines and Healthcare Regulatory Agency (MHRA)

Thalidomide

The **thalidomide** drug was developed, tested and **approved** as a sleeping pill.

It was also found to be effective in relieving morning sickness in pregnant women. But it **hadn't been tested** for this use.

It was **banned** when many women who took the drug gave birth to babies with severe limb abnormalities.

Thalidomide was **re-tested**. It is now used to treat **leprosy**.

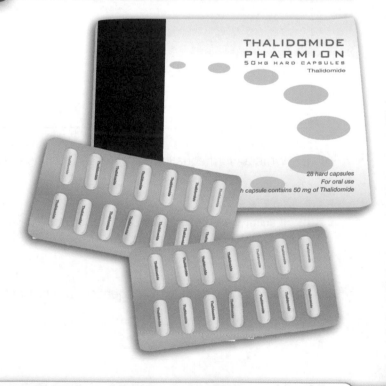

THALIDOMIDE
PHARMION
50MG HARD CAPSULES
Thalidomide

28 hard capsules
For oral use
h capsule contains 50 mg of Thalidomide

Pathogens

Microorganisms that cause **infectious** diseases are called **pathogens**. There are two main types of pathogen that can affect your health:

- **Bacteria**
- **Viruses.**

Bacteria	Viruses
Very small.	Smaller than bacteria.
Reproduce rapidly.	Reproduce rapidly once inside living cells, which they then damage.
Can produce **toxins**, which make you feel ill.	Can produce toxins, which make you feel ill.

Treatment of Disease

Medicines such as **painkillers** (e.g. aspirin and paracetamol) are used to **alleviate the symptoms** of disease. However, painkillers don't kill pathogens.

Antibiotics (e.g. penicillin) are used to **kill the infective bacterial pathogens** inside the body. However, antibiotics **don't kill viral pathogens** that live and reproduce inside cells.

It's difficult to develop drugs that kill viruses without damaging your body's tissues.

Many strains of bacteria, including **MRSA**, have developed **resistance** to antibiotics as a result of **natural selection**.

It is necessary to prevent over-use of antibiotics in order to…

- reduce antibiotic resistance
- save the NHS money.

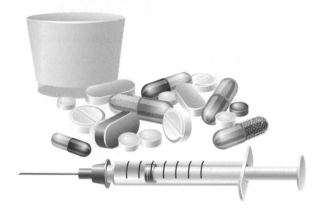

HT Resistance to Antibiotics

The DNA of some bacteria changes (**mutation**). This mutation prevents an antibiotic killing the bacteria, i.e. these bacteria have resistance. They are resistant strains.

Antibiotics kill the bacteria that don't have the mutation, i.e. these bacteria don't have resistance. They are non-resistant strains.

The resistant bacteria survive and reproduce, so the population of resistant bacteria increases.

Nowadays, antibiotics are only used to treat serious infections in order to slow down the rate of development of resistant strains of bacteria.

U3 The Use (and Misuse) of Drugs

Legal and Illegal Drugs

Some people use drugs illegally for **recreation** (pleasure), but they can be very harmful.

- Some recreational drugs are **legal** (alcohol, tobacco)
- Some are **illegal** (e.g. heroin and cocaine, which are very addictive).

Legal drugs (e.g. anti-depressants, amphetamines and barbiturates) have a bigger overall impact on health than illegal drugs, because far more people use them.

Drugs alter chemical processes in the body, so people can become dependent on or **addicted** to them.

Once they are addicted, people will suffer **withdrawal symptoms** if they don't have the drugs.

Withdrawal symptoms may be **psychological** (e.g. paranoia) and **physical** (e.g. sweating and vomiting).

Alcohol and Tobacco

Alcohol contains the chemical **ethanol**. Alcohol can have the following effects:

- It helps people to relax.
- It affects the nervous system, causing reactions to slow down.
- An excess can lead to a lack of self-control, unconsciousness, coma or death.
- It can lead to liver damage or brain damage in the long-term.

Tobacco smoke contains carbon monoxide, nicotine (which is addictive) and **carcinogens**. Tobacco can cause...

- emphysema – alveoli damage caused by coughing
- bronchitis
- arterial and heart disease
- lung cancer.

The carbon monoxide in tobacco smoke reduces the oxygen-carrying capacity of the blood. In pregnant women this can...

- deprive the **fetus** of oxygen
- lead to a **low birth mass**.

Quick Test

1. Give one reason why new drugs are tested in the laboratory.
2. Name three substances found in tobacco.

Pathogens

Pathogens enter our body because of unhygienic conditions or contact with infected persons. Pathogens can enter the body...
- through wounds that break the skin
- through the respiratory system when we breathe in
- by sexual transmission.

Some pathogens that make us feel ill can reproduce rapidly because of the suitable conditions in our body, i.e. warmth, oxygen and a food source. Bacteria produce toxins and viruses damage cells.

Bacteria cause illnesses like TB, tetanus, cholera and typhoid. **Viruses** cause illnesses like influenza, measles, polio, mumps and rubella.

Blood

Blood is a mixture of white and red cells, **platelets**, chemicals and water. White blood cells and platelets help to protect the body.

Platelets contain chemicals that clot the blood. This reduces the amount of blood that leaves the body. Platelets help to form scabs over wounds, which reduces the chance of pathogens entering the body.

White blood cells help to defend against pathogens in several ways.

Vaccination

You can acquire immunity to a particular disease by being **vaccinated** (**immunised**).
1. An inactive / dead pathogen is injected into your body.
2. Your white blood cells produce antibodies to destroy the inactive or dead pathogen.
3. You then have an **acquired immunity** to this particular pathogen because your white blood cells are sensitised to it and will respond to any future infection by producing antibodies quickly.

An example of a **vaccine** is the MMR vaccine used to protect children against measles, mumps and rubella.

Natural Barriers to Pathogens

The risk of catching a disease is reduced by stopping pathogens entering our body. Our body has several natural defences:
- Skin is a physical barrier and is slightly acidic (pH 5.5), which kills many microbes.
- Sweat, tears and ear wax can all destroy microbes.
- Cuts are sealed by blood clots.
- The windpipe (trachea) is lined with mucus, which traps microbes, and has tiny hairs (cilia) that move the mucus up and into the oesophagus.
- Stomach acid kills microbes.

Once pathogens have entered the body, white blood cells can help to destroy them.

There are different types of white blood cells:
- **Phagocytes** can dissolve or engulf and digest pathogens.
- **Lymphocytes** produce antitoxins and antibodies. Antibodies stick to pathogens to identify them and also provide immunity to certain diseases.

Babies can be given immunity by antibodies being passed on...
- through the placenta
- through breast milk.

Quick Test
1. What are pathogens?
2. Describe two ways in which the skin prevents pathogens entering the body.
3. Describe two conditions that increase our risk of catching infectious diseases.
4. Describe three ways in which white blood cells help to destroy pathogens.

Isotopes

An **atom** has a small central nucleus made up of **protons** and **neutrons**. The nucleus is surrounded by **electrons**.

All the atoms of a particular **element** have the same number of protons. However, they can have **different numbers of neutrons**. These are called **isotopes**.

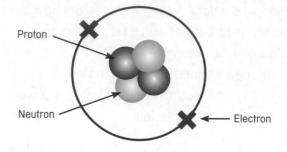

Proton

Neutron

Electron

Radiation

Some substances give out **radiation** all the time. These substances are **radioactive**. Radiation is released from the nucleus of an atom as the result of a change in the atom's structure.

There are three types of radiation:
- **Alpha** (α) particles – a helium nucleus (made up of two protons and two neutrons).
- **Beta** (β) particles – a high-energy electron that is ejected from the nucleus.
- **Gamma** (γ) rays – high-frequency electromagnetic radiation.

When radiation collides with atoms or molecules, it can knock electrons out of their structure, creating a **charged particle** called an **ion**.

Each type of radiation has a different…
- relative **ionising power**
- penetrating power
- range in air.

Particle	Ionising Power	Penetrating Power
Alpha	Strong	Absorbed by a few centimetres of air or thin paper.
Beta	Medium	Passes through air and paper. Absorbed by a few centimetres of aluminium.
Gamma	Weak	Very penetrating. Needs many centimetres of lead or metres of concrete to stop it.

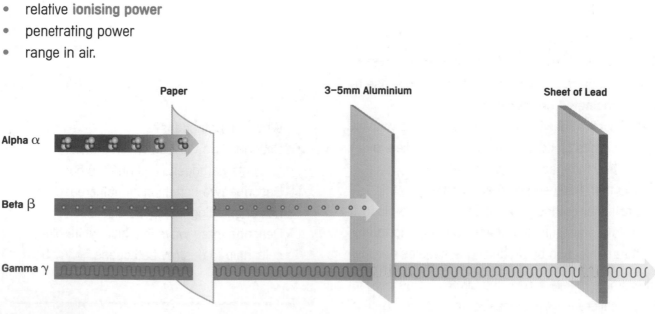

Paper

3–5mm Aluminium

Sheet of Lead

Alpha α

Beta β

Gamma γ

Atom • Element • Isotope • Radiation • Ion • Ionising power

Acute Radiation Dangers

The **ionising power** of alpha, beta and gamma radiation can damage molecules inside healthy living cells. This results in the death of the cell.

Damage to cells in organs can cause **cancer**. The larger the dose of radiation, the greater the risk of cancer.

The damaging effect of radiation depends on whether the source is inside or outside the body.

If the source is **inside the body**…
- alpha causes most damage as it is easily absorbed by cells, causing the most ionisation
- beta and gamma cause less damage as they are less likely to be absorbed by cells.

If the source is **outside the body**…
- alpha cannot penetrate the body; it is stopped by the skin
- beta and gamma can penetrate the body to reach the cells of organs, where they are absorbed.

Inside the Body

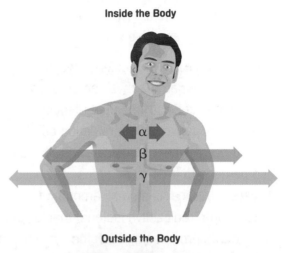

Outside the Body

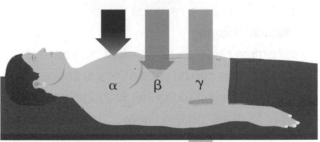

Background Radiation

Background radiation is not harmful to our health as it occurs in very small doses.

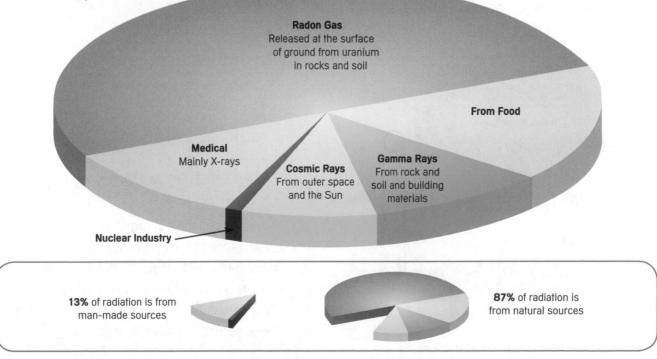

Radon Gas
Released at the surface of ground from uranium in rocks and soil

From Food

Medical
Mainly X-rays

Cosmic Rays
From outer space and the Sun

Gamma Rays
From rock and soil and building materials

Nuclear Industry

13% of radiation is from man-made sources

87% of radiation is from natural sources

Common Uses of Radiation

Gamma rays and X-rays are types of electromagnetic radiation. They are transverse waves.

This radiation can be used in different ways:

- **Sterilisation** – gamma rays can be used to sterilise medical instruments and food.
- **Treating cancer** – gamma radiation can be used to destroy cancerous cells. X-rays can also be used to treat some types of cancer.
- **Tracers** – a tracer is a small amount of a radioisotope (radioactive isotope) that is put into the body. Its progress through the system can be traced using a radiation detector.
- **Medical imaging**
  - gamma rays can be detected by a gamma camera; this can be used to detect cancer.
  - X-rays pass through soft tissue so can be used to detect broken bones.

Working with high energy radiation like gamma and X-rays can be harmful. People who work with radiation wear **film badges** that are checked regularly to measure their exposure to radiation.

Film badges contain a material, sometimes photographic film, which changes when exposed to radiation.

Treating Cancer

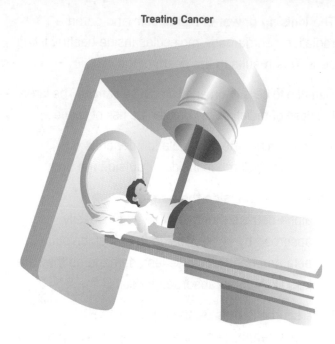

X-rays

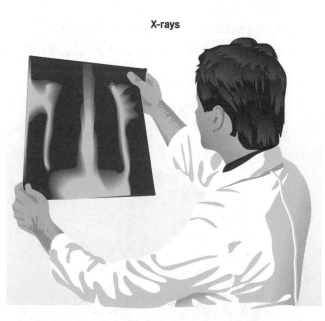

Film Badge

Open window that contains photographic film

Quick Test

1. Name three types of radiation.
2. Which type of radiation causes most damage to cells when it's inside the body?
3. Give two uses of X-rays in hospitals.
4. People who work with radiation are checked for exposure. How is this done?

State Symbols

State symbols are used in equations. The symbols are **(s)** **solid**, **(l)** **liquid**, **(g)** **gas** and **(aq)** **aqueous**.

An **aqueous solution** is produced when a substance is **dissolved in water**. This liquid is called an **electrolyte**.

Ions are charged particles that are formed when atoms gain or lose electrons.

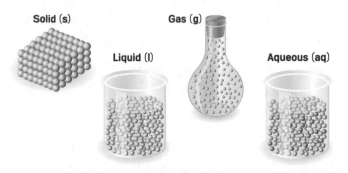

Solid (s) Gas (g)

Liquid (l) Aqueous (aq)

Electrolysis

Electrolysis is the **breaking down** of a compound containing **ions** into its **elements** using an **electrical** current.

Ionic substances are chemical compounds that allow an **electric current** to flow through them when they are…

- molten
- dissolved in water.

These compounds contain **negative** and **positive** ions. During electrolysis…

- **negatively charged ions** move to the **positive electrode** (anode)
- **positively charged ions** move to the **negative electrode** (cathode).

This moving of electrons forms electrically **neutral** atoms or molecules that are then released.

If there is a **mixture of ions** in the solution, the products formed depend on the reactivity of the elements involved.

For example, in the electrolysis of **copper chloride solution,** the substances released are…

- copper at the negative electrode
- chlorine gas at the positive electrode.

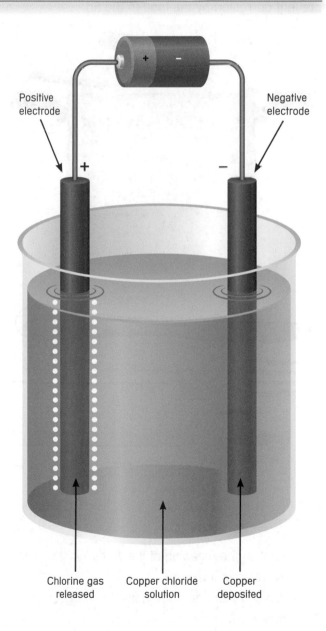

Positive electrode

Negative electrode

Chlorine gas released Copper chloride solution Copper deposited

Electroplating

When metals or substances that conduct electricity are coated with a metal, it is called **electroplating**.

Some metals are electroplated to…
- prevent corrosion
- make them look more attractive.

Saucepans, cutlery and taps are electroplated for these two reasons. Food tins are coated with tin to prevent corrosion by the food in the tin.

Some jewellery made from cheaper metal is coated with silver and gold to make it look more attractive.

Jewellery made from nickel can cause allergies. It is electroplated with precious metals to prevent allergic reactions.

The item being plated is used as the **cathode**. The metal to be used for plating (e.g. silver) is the **anode**:

1. Both the item and the coating metal are placed in an aqueous solution of the metal.
2. When the electric current flows through the solution, the metal is deposited on the item being plated.
3. Metal ions from the anode enter the solution.
4. At the negative electrode (cathode) metal (e.g. silver) is deposited.

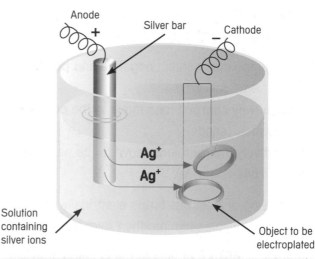

Anode · Silver bar · Cathode · Ag^+ · Ag^+ · Solution containing silver ions · Object to be electroplated

HT The processes that take place at the electrodes can be represented by equations. At the positive electrode (anode), silver atoms become silver ions.

$$\text{Anode} \quad Ag \longrightarrow Ag^+ + e^-$$

At the negative electrode (cathode) silver metal is deposited.

$$\text{Cathode} \quad Ag^+ + e^- \longrightarrow Ag$$

Quick Test

1. Explain the meaning of the word 'electrolysis'.
2. What is an ion?
3. Which electrode do negatively charged ions move to?
4. When electroplating an item, which electrode should it be used as?

Nanoparticles and Nanostructures

Nanoscience is the study of very small structures. The structures are 1–100 nanometres in size, roughly in the order of a few hundred atoms.

One **nanometre** is 0.000 000 001m (one billionth of a metre) and is written as 1nm or 1×10^{-9}m.

Nanoparticles are tiny, tiny particles that can combine to form structures called **nanostructures**.

Nanostructures can be manipulated so materials can be developed that have new and specific properties.

The **properties** of nanoparticles are different to the properties of the **same materials in bulk**.

For example...
* electrons can move through an insulating layer of atoms
* nanoparticles are more sensitive to light, heat and magnetism
* nanoparticles have a high surface area in relation to their volume.

Human Hair	Virus / Small Bacterium	Atoms and Small Molecules
0.000 01 metre = 1×10^{-5}m	0.000 000 1 metre = 1×10^{-7}m	0.000 000 001 metre = 1×10^{-9}m
Can be seen using a microscope	Can be seen using an electron microscope	Can be seen using an electron microscope

Nanocomposites

Other materials can be added to plastics to make **nanocomposite** materials. In comparison to plastics, nanocomposites can be...
* stronger
* stiffer
* lighter.

The characteristics of nanocomposites can be seen by looking at the nanostructures formed by the nanoparticles.

Nanocomposites are already being used or developed for many industries, for example...
* the car industry
* medical and dental applications
* new computers
* highly selective sensors
* new coatings and sunscreens
* stronger and lighter construction materials
* product-specific catalysts.

Smart materials are a type of nanostructure that can be designed to...
* have specific properties on a nanoscopic scale
* behave in a certain way when subjected to certain conditions.

New Products

Smart paint is a type of coating that heals its own scratches (i.e. it's self healing) when exposed to sunlight.

It can be used for cars, aeroplanes, electronic items and anything that might be scratched.

Superconductors are substances whose resistance becomes almost zero at low temperatures. This reduces energy transfers.

Superconductors are used to make powerful electromagnets that are used...

*   in MRI scanners
*   for magnetic levitation, e.g. Maglev trains that move using basic magnetic principles and don't have an engine that uses fossil fuels.

Smart materials are substances that are able to change their properties in response to the environment. Applications of smart materials include dental braces, spectacle frames, shrink wrap packaging and wound dressings.

Chromic materials are materials that are able to change colour when conditions change.

Thermochromic materials change colour as the temperature changes. These can be used in thermometer strips for...

*   medical use
*   fish tanks.

Thermochromic ink can be used on food labels for food that needs to be kept cold.

Photochromic materials change colour as light intensity changes. These materials can be used in...

*   spectacle lenses
*   light detectors
*   optical switches
*   rear view mirrors
*   light intensity meters
*   necklaces that change colour after exposure to too much sunlight.

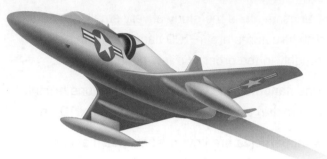

Shrink Wrap Packaging

Thermometer Strip

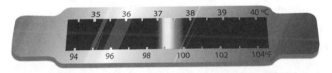

Quick Test

1. What are nanoparticles?
2. What is smart paint?
3. Give one use of superconductors.
4. Name two chromic materials and the conditions that they are able to respond to.

Selective Breeding and Genetic Engineering

Selective Breeding

Selective breeding is used to produce animals with certain characteristics:

1. Parents are selected that have the desired characteristics.
2. They are crossed to produce offspring.
3. The offspring with the desired characteristics are then crossed.
4. This continues for several generations.

Reproducing Plants

Plants can reproduce **asexually**. Offspring produced asexually are **clones**. Many plants naturally reproduce asexually, e.g. spider plants and strawberry plants.

Many plants can be reproduced asexually **by artificial means**. For example, you can take **cuttings** from a plant with desired characteristics to produce clones quickly and cheaply.

Spider Plant Stolons

Stolon – a rooting side branch   New individual established   Now independent

Plant Cuttings

Cloning

One modern cloning technique is **tissue culture**.

Small groups of cells are taken from part of a plant or animal and grown in medium containing nutrients and hormones. The offspring are **genetically identical to the parent**.

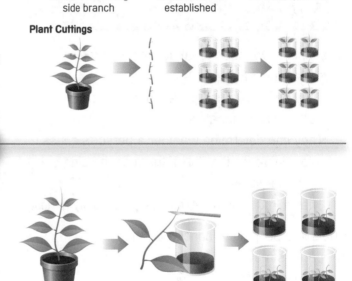

Genetic Modification

Genetic modification (genetic engineering) involves transferring genetic material from one organism to another.

Genes from the chromosomes of humans and other organisms are **cut out** using **enzymes**. Then they are transferred to cells of other organisms.

In animals and plants, genes are often transferred at an early stage of their development so that the organism develops with **desired characteristics**.

Reasons for altering an organism's genetic make-up are…

- to improve crop yield
- to improve resistance to pests or herbicides
- to extend the shelf-life of fast-ripening crops
- to harness the cell chemistry of an organism so that it produces a required substance, e.g. production of human **insulin**
- to reduce the effect of damaged genes.

Insulin Production

Insulin is the **hormone** that helps to control the level of glucose in your blood.

People who have diabetes can't produce enough insulin and often need to inject it.

Human insulin can be produced by genetic engineering.

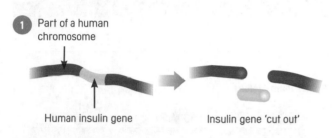

1 Part of a human chromosome

Human insulin gene

Insulin gene 'cut out'

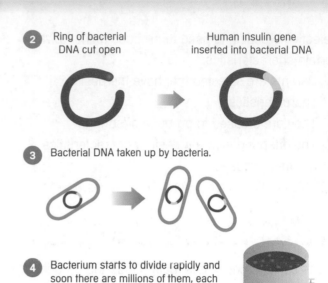

2 Ring of bacterial DNA cut open

Human insulin gene inserted into bacterial DNA

3 Bacterial DNA taken up by bacteria.

4 Bacterium starts to divide rapidly and soon there are millions of them, each with instructions to make insulin.

VAT

The Great Genetic Debate

Scientists have made great advances in their understanding of genes:

- They have identified genes that control certain characteristics.
- They can determine whether a person's genes might increase the risk of them contracting a particular illness.
- They might soon be able to 'remove' faulty genes.

But, some people are concerned that…

- parents may want to choose the genetic make-up of their children
- unborn babies may be aborted if their genetic make-up is faulty
- insurance companies will genetically screen applicants and refuse to insure people who have an increased genetic risk of illness or disease.

DESIGNER BABY SERVICE

Choose the eyes you want **Choose the lips you want**

CONTACT DB SERVICES FOR PRICES

It is possible to buy a gift voucher for genetic screening in America. This provides all your genetic information, including your chance of developing a genetically determined disease.

Quick Test

1 What are produced when plants reproduce asexually?
2 Describe how a tissue culture of a carrot plant could be made.
3 What is genetic modification?
4 Name a hormone used to control glucose levels that can be made using genetic engineering.

The Population Explosion

The standard of living for most people has improved a lot over the past 50 years. The **human population** is now **increasing** **exponentially** (i.e. with accelerating speed).

This rapid increase in the human population means that...

- raw materials, including **non-renewable energy resources**, are being used up quickly
- more and more **waste** is being produced (so more landfill sites are needed)
- improper handling of waste is causing **pollution**
- there is less land available for plants and animals due to...
    - farming
    - quarrying
    - building
    - dumping waste.

Pollution

Human activities may pollute...

- **water** – with sewage, fertilisers or toxic chemicals
- **air** – with smoke and gases, such as carbon dioxide, sulfur dioxide and oxides of nitrogen that contribute to acid rain
- **land** – with toxic chemicals like **pesticides** and **herbicides** that may be washed from land into water.

Unless waste is properly handled and stored, more pollution will be caused.

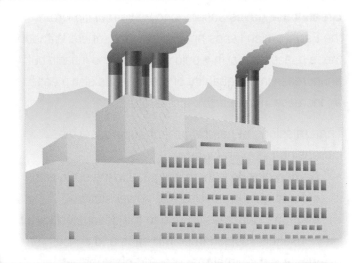

Indicators of Pollution

Living organisms can be used as **indicators of pollution**. For example...

- the types of **lichen** species present can indicate air pollution (lichens are a blend of algae and fungus)
- the presence or absence of invertebrate animals can indicate water pollution, e.g. freshwater shrimp survive only in unpolluted water.

Lichen

Lichen indicates air pollution

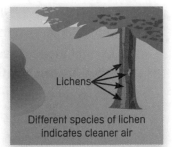

Lichens

Different species of lichen indicates cleaner air

Deforestation

Deforestation involves the large-scale cutting down of trees...
* for timber
* to provide land for agricultural use.

Deforestation has occurred in many tropical areas with **devastating consequences** for the environment.

Deforestation has...
* increased the amount of carbon dioxide (CO_2) released into the atmosphere (due to burning of wood and decay of wood by microorganisms)
* reduced the rate at which carbon dioxide is removed from the atmosphere by **photosynthesis**.

Deforestation reduces **biodiversity** and results in the loss of different species that could be an important part of an ecosystem.

Global Warming

Greenhouse gases absorb long wave radiation from the Earth, which keeps heat in the atmosphere. Without these gases the Earth would be too cold to inhabit, but an increase in the quantity of these gases also means an increase in the Earth's temperature.

The emissions of natural greenhouse gases are increasing in several ways:
* **Carbon dioxide** is increasing due to the combustion of fossil fuels in vehicles and power stations.
* **Methane** is increasing due to the decomposition of household rubbish in landfill sites and agricultural waste, e.g. animal manure and plant waste.
* Methane is also produced from waterlogged soil where there is very little air, e.g. rice fields and swamps. Bacteria in the waterlogged soil produce methane. These bacteria are also found in the gut of farm animals (e.g. cows and sheep).
* **Nitrous oxide** concentrations are increasing from vehicle exhausts and power stations and from increasing use of nitrogen-based fertilisers.

The **Kyoto agreement** is an agreement between many industrialised countries to reduce the emissions of greenhouse gases that they produce by 2012 relative to their emissions in 1990. Some countries refused to join the agreement, including Australia and the USA.

The Kyoto protocol that followed the agreement set legally binding targets for the thirty-seven member countries.

Eutrophication

An increase in nitrates and phosphates from fertilisers and sewage in freshwater lakes and rivers causes **eutrophication**. It damages ecosystems.

1. Fertilisers are chemicals used to increase the amount of crop growth. These chemicals can be washed off the land or leached into underground water supplies that then flow into lakes and rivers.

2. When artificial fertilisers enter lakes and rivers they help algae and water plants to grow.

3. The algae cover the surface of the water and prevents light reaching the plants growing below the surface.

4. The plants can't photosynthesise, so they die.

5. Bacteria that decompose the dead plants use up oxygen in the water for respiration.

6. As the oxygen concentration in the water decreases, more animals die and they are then also decomposed by bacteria.

7. There are very few living plants and animals left in the water. This is eutrophication.

Nitrates cause excessive algal growth, which blocks off sunlight to other plants

The other plants can't photosynthesise so they die

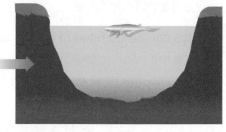

Bacteria involved in rotting processes use up oxygen and the water can't support life

Sustainable Development

Sustainable **development** means improving the quality of life on Earth without compromising future generations. It needs to be considered at…

- **local** levels
- **regional** levels
- **global** levels.

Sustainable development is concerned with three related issues…

- **economic** development
- **social** development
- **environmental** protection.

Sustainable resources are resources that can be **maintained** in the long term at a level that allows appropriate consumption / use by people.

This often requires **limiting exploitation** by using **quotas**, or by ensuring the resources are **replenished / restocked**.

Example 1

The UK has one of the largest sea fishing industries in Europe.

To ensure the industry can continue and fish stocks can be conserved…

- quotas are set to prevent over-fishing
- the mesh size of nets has been increased to prevent young fish being caught before they reach breeding age.

Example 2

Scandinavia uses a lot of pine wood to make furniture and paper, and to provide energy.

To ensure the long-term economic viability of pine-related industries, companies restock the pine forests by planting a new sapling for each tree they cut down.

Disposing of Plastics

Plastic is a versatile material. It is cheap and easy to produce, but this means a lot of plastic waste is generated.

There are various ways of **disposing of plastics**. Unfortunately some of them have an impact on the environment.

Burning plastics produces **air pollution** and releases carbon dioxide, which contributes to **global warming**.

Some plastics are not safe to burn at all because they produce toxic fumes.

Plastics can be dumped in **landfill sites**, but most plastics are **non-biodegradable**. This means that microorganisms have no effect on them, so they will not **decompose** and rot away.

The use of landfill sites means that plastic waste builds up.

Research is currently being carried out into the development of **biodegradable plastics**. Biodegradable plastic bags are already being used.

HT Degradation of Plastics

Water-soluble Plastics

Some plastics are **biodegradable**, which means that they can be broken down by microorganisms.

For example, Polyvinyl Alcohol (PVOH) and Ethylene Vinyl Alcohol (EVOH) are biodegradable. These plastics are also soluble in water in certain conditions.

They can be made into plastic films to cover food and for shopping bags.

Biodegradable Plastics

Some plastics are being made that can be broken down by…

- light, e.g. **photo-degradable plastics** degrade after being exposed to sunlight for a long time
- the use of a chemical additive and microbial action, where an additive is used to start the breakdown of the plastic so that microbes can then complete the degradation, e.g. **oxo-degradable plastics**.

Quick Test

1. Name two types of organisms that can be used as pollution indicators.
2. Name two greenhouse gases.
3. Describe two ways in which fertilisers can enter lakes and rivers.
4. Only a few plastics are 'biodegradable'. Explain the meaning of biodegradable.

Conduction

Conduction is the **transfer** of heat **energy** without the substance itself moving.

The structure of **metals** makes them good **conductors** of heat:

1. As a metal becomes hotter, its tightly packed particles gain more kinetic energy and vibrate.
2. This energy is transferred to cooler parts of the metal by delocalised electrons, which move freely through the metal, colliding with particles and other electrons.

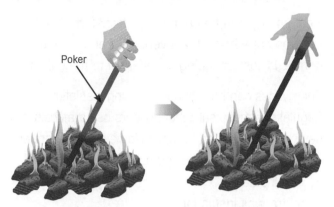

Poker

Heat energy is conducted up the poker as the hotter parts transfer energy to the colder parts

Convection

Convection is the transfer of heat energy through **movement**.

Convection occurs in **liquids** and **gases** and creates **convection currents**:

1. The particles in the liquid or gas that are nearest the heat source move faster, causing the substance to expand and become less dense than in the colder parts.
2. The warm liquid or gas will rise up. The liquid or gas cools, becomes denser and sinks. The colder, denser liquid or gas moves into the space created (close to the heat source), replacing the liquid or gas that has risen.

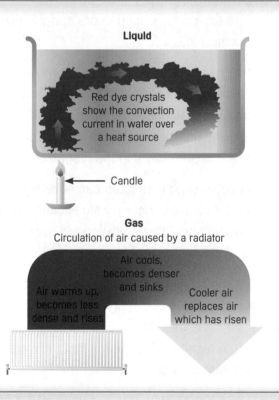

Liquid

Red dye crystals show the convection current in water over a heat source

Candle

Gas
Circulation of air caused by a radiator

Air cools, becomes denser and sinks

Air warms up, becomes less dense and rises

Cooler air replaces air which has risen

Radiation

Thermal (**infrared**) **radiation** is the transfer of heat (thermal) energy by **electromagnetic waves**. No particles of matter are involved.

All objects emit and absorb thermal radiation. The hotter the object, the more energy it radiates. The amount of radiation an object gives out or takes in depends on its **surface**, **shape** and **dimensions**.

An object will emit or absorb energy faster if there's a big difference in temperature between it and its surroundings.

Under similar conditions, different materials transfer heat at different rates. At the same temperature, dark matt surfaces…

- **emit** more radiation than light shiny surfaces
- **absorb** more radiation than light shiny surfaces.

Reducing Energy Loss at Home

Heat energy can be changed or transferred, but it can't be created or destroyed. Heat energy in our homes is 'lost' by being transferred to the outside.

Energy loss can be reduced by using insulation. Regulations make sure that builders use materials that have better insulation to minimise heat loss. There are various methods of reducing heat transfer including...

- loft insulation
- extra wall insulation
- hot water tank jackets
- thermostatic controls to maintain a constant temperature
- draught excluders around doors and windows
- double glazing.

Each energy-saving measure costs money and some are more efficient than others at reducing energy loss. **Payback time** for an energy-saving measure is the number of years it takes to save the money that it cost to buy it in the first place.

Some energy-saving measures can have high efficiency. This means that they significantly reduce energy loss, but could be so expensive to buy that it isn't cost effective for most people to have.

U-value is the measure of the rate of heat loss through a material. All building materials are given a U-value, which is used to choose the best material for the job. A high U-value means that the material has a high rate of heat loss.

For example, if a window has a U-value of 2.2 and a wall has a U-value of 0.35, the wall has a lower rate of heat loss.

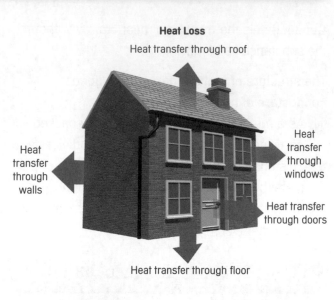

Heat Loss

Heat transfer through roof

Heat transfer through walls

Heat transfer through windows

Heat transfer through doors

Heat transfer through floor

Loft Insulation

Double Glazing

Wall Insulation

Hot Water Tank Jacket

Controlling Pollution

Common pollutants in the home are...
- dust
- mould and spores
- pollen
- smoke
- fumes from household products.

Symptoms of exposure to high levels of indoor pollution include...
- asthma
- headaches
- tiredness
- dizziness
- nausea
- itchy nose
- sore throat.

Chemicals used in our homes can be harmful. Labels on the bottles give advice about...
- how to use them safely
- how to reduce the risks during use
- what to do if an accident happens.

Heating Boilers

Heating boilers and gas fires need enough oxygen from the air to burn fossil fuels completely. Incomplete combustion of fuel due to a restricted air supply...
- releases toxic carbon monoxide, which is odourless
- releases soot
- causes a lower energy output.

Carbon monoxide detectors can be used near boilers and gas fires to keep people safe.

Reducing draughts does reduce heat loss, but good ventilation in houses will reduce the risk of incomplete combustion.

Radon

Radon gas is a radioactive gas that can cause cancer. It can be a pollutant in houses that are built on rocks or soil that contain radium or uranium

Houses built on these rocks are designed to remove radon gas before it can cause harm.

Quick Test

1. Name three ways heat energy can be transferred.
2. Which two types of heat transfer need particles to transfer the heat?
3. Can energy be lost?
4. What is a U-value?
5. Identify five substances that are common pollutants in our homes.
6. Give one reason why some chemicals used in the home need hazard symbols on their labels.
7. Name a toxic substance produced when incomplete combustion of fossil fuels take place.

1 a) What is a drug?

.... **(1 mark)**

b) Why is it necessary to prevent overuse of antibiotics?

.... **(1 mark)**

2 a) Complete the sentences using words from the box.

beta	gamma	radiation	wood	many	one	paper

All radioactive substances emit .... radiation is able to

penetrate paper but is stopped by a few millimetres of aluminium. Gamma radiation is the most

penetrating and is only stopped by .... centimetres of lead. Alpha radiation is stopped by

a thin sheet of .... **(4 marks)**

b) 'The damaging effect of alpha, beta and gamma radiation depends on whether the radioactive source is inside or outside the body'. Explain what this statement means.

....

....

....

....

....

....

.... **(6 marks)**

3 Sulfur dioxide is an acidic gas that reduces the growth of plants. Sodium metabisulfate is a chemical that decomposes to release sulfur dioxide. This chemical can be used to investigate the effect of sulfur dioxide on the growth of plants.

Stephen and Nate investigated the effect of different concentrations of sulfur dioxide on the growth of cress plants. They recorded their results in the table opposite.

Concentration of sulfur dioxide	Plant growth
1	2
2	7
3	1

a) The headings of the table are not complete. What should they have included in their headings to make their results clear?

.... **(1 mark)**

b) They only tested three concentrations of sulfur dioxide. How many should they have tested and why?

....

.... **(2 marks)**

c) What evidence from the table suggests that their results may not be reliable?

....

.... **(2 marks)**

4 Air pollution affects the growth of many species of organisms. Lichens are affected by air pollution and will only grow in specific places. Scientists investigated the number of lichen species found in two areas. The table below shows the results of a survey of lichens growing on limestone walls. They found six different species of lichen and identified them by number.

	Species of lichen present
Area A	Species 1
	Species 3
	Species 4
Area B	Species 1
	Species 2
	Species 5
	Species 6

a) The scientists concluded that area A was less polluted than area B. Explain how it was possible for the scientists to reach this conclusion using the results from the table.

....

....

....

.... **(3 marks)**

b) Species 1 was found in both areas. Suggest an explanation for this result.

.... **(1 mark)**

Answers

Quick Test Answers

Page 13
1. Reflecting, refracting, radio and X-ray.
2. By using a lens.
3. The images produced aren't affected by atmospheric interference.
4. The wavelengths from the wave source are longer than if it wasn't moving.

Page 16
1. Crust, mantle and core.
2. Convection currents are created by intense heat from radioactive decay.
3. They may cause earthquakes and tsunamis.

Page 19
1. Carbon dioxide levels have decreased. Nitrogen and oxygen levels have increased.
2. Burning fossil fuels.
3. A rise in global temperature could cause substantial climate change and cause sea levels to rise.

Page 22
1. Protons, neutrons and electrons.
2. The number of neutrons is equal to the mass number minus the atomic number.
3. They can share electrons or give electrons or take electrons.
4. Chemical formulae show the different elements in a compound and the number of atoms of each element in a compound.

Page 28
1 Materials can be removed from the ground by mining and quarrying.
2. The process is called fractional distillation.
3. Extraction uses large amounts of energy.
4. Carbon and carbon monoxide
5. iron oxide + carbon monoxide $\longrightarrow$ iron + carbon dioxide

Page 30
1. Isotopes are atoms of the same element that have different numbers of neutrons.
2. $\dfrac{\text{Relative mass of element in the compound}}{\text{Relative formula mass } (M_r)} \times 100$

Page 32
1. It makes the reaction economical and ensures sustainable development.
2. $\dfrac{\text{Yield from reaction}}{\text{Maximum theoretical yield}} \times 100$

Page 36
1. They look for similar characteristics.
2. Plants compete for light, water and nutrients from the soil.
3. Plants have water storage tissue and extensive root systems.

Page 38
1. The Sun is the source of energy.
2. Materials and energy are lost in body waste and released through respiration.
3. Oxygen, warmth and moisture
4. Decay

Page 39
1. The two processes that are balanced are the removal and return of carbon.
2. The process that releases carbon dioxide is respiration.
3. Carbon is stored in fossil fuels and in the shells and bones of sea animals.

Exam Practice Answers
1. **a)** crust **b)** mantle **c)** core
2. **a)** **Any two from:** Closely matching coastlines; Similar patterns of rocks; Contain fossils of the same plants and animals.
   b) **In any order:** Plates slide past each other; Plates move towards each other; Plates move away from each other.
   c) **Any two from:** Earthquakes; Volcanic eruptions; Tsunamis.
3. A is gas 3; B is gas 4; C is gas 1; D is gas 2
4. **a)** The idea that all living things evolved from simple life forms; that first developed three billion years ago.
   b) **i)** The remains of plants or animals; from many years ago that are found in rock.
   ii) Provide evidence of how organisms have changed over time.
5. competition; environment; predators; disease
6. A – blackbird; B – ladybird; C – greenfly; D – oak tree

Quick Test Answers

Page 45
1. Motor, sensory and relay neurone.
2. Temperature, blood sugar (glucose) levels, water content and ion content.
3. The hormone is called glucagon.
4. The thermoregulatory centre monitors and controls the body temperature.

Page 47
1. A hydrogen ion reacts with a hydroxyl ion to produce water.
2. An antacid is a base.
3. Hydrogen
4. You would use sulfuric acid to make a metal sulfate.

Page 52
1. Respiration.
2. Genetic and environmental factors.
3. Human body cells contain 46 chromosomes.
4. An allele is a different form of a gene.

Page 57
1. Copper is a good conductor of electricity.
2. Ceramics are made by heating inorganic materials.

3. GRP (glass-reinforced polyester) or fibreglass; Carbon fibre-reinforced epoxy resin; Plasticised PVC.
4. They have at least one carbon–carbon double bond.

Page 58
1. Hydrogen and carbon.
2. The carbon atoms are joined by single bonds.
3. Carbon dioxide and water.

Page 63
1. Coal, oil and gas.
2. Fossil fuels are burned to release thermal energy. The thermal energy is used to boil water to produce steam. The steam drives turbines that are attached to electrical generators.
3. Transformers are used to change the voltage of the electricity passing through the cables.
4. It can't be replaced in a lifetime.
5. **Advantages, accept two from:** Fuel costs relatively low; Stations flexible in meeting demand; Stations can be situated in sparsely populated areas; No CO_2 or SO_2 produced.
   Disadvantages, accept two from: A long start-up time; Building and decommissioning is expensive; Problems with radioactive waste; Problems with radioactive leaks.

Answers

Unit 2 (Cont.)

6. **Advantages, accept two from:** No fuel costs during operation; Very little chemical pollution; Low maintenance.
    Disadvantages, accept two from: Visual pollution; Uses up lot of land/space; High initial capital outlay; Generally only produces small amounts of electricity; Generally unreliable as they are weather-dependent.

Page 67

1. Efficiency = $\dfrac{\text{Useful energy transferred by the device}}{\text{Total energy supplied to device}}$ x 100%

2. The amount of energy transformed depends on how long the device is switched on and how fast the appliance can transform energy.

3. Number of kilowatt-hours x Cost per kilowatt-hour.

4. Radio waves.

Exam Practice Answers

1. **a)** A 3; B 2; C 4; D 1
    b) Synapse
2. **a)** **i)** potassium chloride **ii)** sulfuric acid **iii)** sodium hydroxide
    iv) barium sulfate + zinc chloride
    b) barium sulfate
3. **a)** A = cell membrane; B = nucleus; C = cytoplasm

    b) This is where most energy is released during respiration
    c) Cell membrane
4. **a)** Differences between individuals of the same species
    b) Genetic; Environmental
    c) Brown-eyed mother is heterozygous and so has the recessive b allele for blue eyes; Blue-eyed father must be homozygous for the recessive b allele; 50% chance of blue-eyed offspring.

    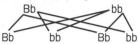

5. A 1; B 2; C 3
6. **a)**

    H—C—C—H structure with H atoms above and below each carbon

    (1 mark for the atoms; 1 mark for all single bonds)

    b) **i)** ethane + oxygen $\longrightarrow$ carbon dioxide and water
    (1 mark for reactants, 1 mark for products)
    ii) sulfur dioxide (SO_2)
    iii) $2C_2H_6 + 7O_2 \longrightarrow 4CO_2 + 6H_2O$

Unit 3

Quick Test Answers

Page 72

1. New drugs are tested in the laboratory to find out if they are toxic.
2. Carbon monoxide, nicotine and carcinogens.

Page 73

1. Pathogens are microorganisms that cause infectious diseases.
2. The skin prevents pathogens from entering the body by being slightly acidic and acting as a barrier.
3. If we are in unhygienic conditions; If we are in contact with infectious diseases.
4. Dissolving or engulfing and digesting them; Producing antitoxins to neutralise toxins; Producing antibodies to identify pathogens and provide immunity.

Page 76

1. Alpha, beta and gamma.
2. Alpha radiation causes most damage when inside the body.
3. X-rays can be used to detect broken bones and treat cancer.
4. They wear film badges that measure their exposure.

Page 78

1. Electrolysis is the breaking down of a compound containing ions into its elements using an electric current.
2. An ion is a charged particle.
3. Negatively charged ions move to the positive electrode (anode).
4. The item should be used as the negative electrode (cathode).

Page 80

1. Nanoparticles are tiny, tiny particles that can combine to form structures called nanostructures.
2. Smart paint is a coating that can repair itself when exposed to sunlight.
3. Superconductors can be used to make electromagnets.
4. Thermochromic materials respond to changes in temperature; Photochromic materials respond to changes in light intensity.

Page 82

1. When plants reproduce asexually they produce clones.
2. A tissue culture could be made by taking a small group of cells from part of the carrot and growing it in soil containing nutrients and hormones.
3. Genetic modification involves transferring genetic material from one organism to another.

4. Insulin can be made using genetic engineering.

Page 86

1. Lichens and invertebrates can be used as pollution indicators.
2. Carbon dioxide and methane.
3. They can be washed off the ground or leach through the ground into underground water supplies.
4. Biodegradable means they can be broken down by microorganisms.

Page 89

1. Heat can be transferred by conduction, convection and radiation.
2. Conduction and convection need particles.
3. Energy can't be lost, it can only be transferred or transformed.
4. The U-value is a measure of the rate of heat transfer through a material.
5. Dust, mould and spores, pollen, smoke, fumes from household products.
6. Chemicals used in the home can be harmful.
7. Carbon monoxide.

Exam Practice Answers

1. **a)** A drug is a chemical that alters the way the body works.
    b) In order to prevent the increase in the number of antibiotic-resistant bacteria; Save the NHS money.
2. **a)** radiation; beta; many; paper
    b) **Inside the body, any three from:** Alpha is most dangerous; Easily absorbed; Alpha causes most ionisation; Beta / gamma causes less damage; Less likely to be absorbed
    Outside the body, any three from: Alpha cannot penetrate the skin; Alpha is least harmful; Beta / gamma more harmful; Penetrate body to reach cells; Beta / gamma easily absorbed
3. **a)** The units are missing.
    b) At least 5; so they could see a trend.
    c) **Any two from:** No repeats; No control; Concentration 2 gives an odd result.
4. **a)** Different species were found in both places; The species of lichens present depends on the pollution present; They knew which species usually grow in polluted areas and which didn't.
    b) It was unaffected by the pollution.

Glossary

Absorption – a substance's ability to absorb energy

Acid – a compound that has a pH value lower than 7

Adaptation – the gradual change of a particular organism over generations to become better suited to its environment

Alkali – a compound that has a pH value higher than 7

Alkane – a saturated hydrocarbon with the general formula C_nH_{2n+2}

Allele – an alternative form of a gene

Alloy – a specific mixture of two or more metals, e.g. brass, or a mixture of a metal and a non-metal

Antibiotic – medication used to kill bacterial pathogens inside the body

Atom – the smallest part of an element that can enter into a chemical reaction

Atom economy – a measure of the amount of reactants in a chemical reaction that end up as useful products; usually expressed as a percentage

Atomic number – the number of protons in the nucleus of an atom

Biodiversity – the variety of species of living organisms and the ecosystems in which they live

Biomass – the mass of a plant or animal minus the water content

Carbon cycle – the constant recycling of carbon by the processes of life, death and decay

Carcinogen – a substance that causes cancer

Chemical formula – a way of showing the elements and atoms that are present in molecules of a substance

Chemical reaction – a process in which one or more substances are changed into others

Chromosome – long structure found in the nucleus of all cells containing DNA

Clone – a genetically identical offspring of an organism

Community – all the living organisms in an area or habitat

Composite – made by combining two or more materials

Compound – a substance consisting of two or more elements chemically combined together

Conduction – the transfer of heat without the substance itself moving

Conductor – a substance that readily transfers heat or electricity

Convection – the transfer of heat energy though the movement of the substance

Covalent bond – a bond between two atoms in which both atoms share one or more electrons

Crude oil – a liquid mixture found in rocks, which contains hydrocarbons

Current – the flow of electric charge through a conductor

Decay – to rot or decompose

Decompose – to break down

Deforestation – the destruction of forests by cutting down trees

Detritus – organic material formed from dead and decomposing plants and animals

Differentiation – cells become specialised

Distillation – a process of separating a mixture by boiling a liquid to evaporate it and condensing its vapours

DNA – nucleic acid that contains the genetic information carried by every cell

Drug – a chemical substance that alters the way the body works

Effector – the part of the body, e.g. a muscle or gland, that produces a response to a stimulus

Efficiency – the ratio of energy output to energy input, expressed as a percentage

Electrode – a piece of metal or carbon that allows electric current to enter and leave a solution during electrolysis

Electrolysis – the process by which an electric current causes a solution to undergo chemical decomposition

Electromagnetic spectrum – a continuous arrangement that displays electromagnetic waves in order of increasing frequency

Electron – a negatively charged subatomic particle that orbits the nucleus

Element – a substance that consists of only one type of atom

Energy – the capacity of a physical system to do work; measured in joules (J)

Environment – the conditions around an organism

Enzyme – a protein catalyst that alters the rate of a particular biochemical reaction

Evidence – observation, measurements and data collected and subjected to some form of validation

Evolution – the natural change of a species or animal over a period of time

Exponentially – with accelerating speed

Extinct – a species that has died out

Fetus – an unborn animal / human baby

Food chain – the feeding relationship between organisms

Fossil – the remains of animals / plants preserved in rock

Fossil fuel – fuel formed in the ground, over millions of years, from the remains of dead plants and animals

Fractional distillation – the process used to separate mixtures, e.g. crude oil, into groups of hydrocarbons (called fractions) whose molecules have a similar number of carbon atoms

Fuel – a substance that releases heat or energy when combined with oxygen

Gamete – sex cell

Gene – a small section of DNA in a chromosome that determines a particular characteristic

Gland – an organ in an animal's body that secretes substances

Global warming – the increase in the average temperature on Earth due to a rise in the levels of greenhouse gases in the atmosphere

Gravitropism – a plant's growth response to gravity

Greenhouse effect – the process by which the Earth is kept warm by heat reflecting back to the Earth

Herbicide – a toxic substance that is used to destroy unwanted vegetation

Hormone – a regulatory chemical that stimulates cells or tissues into action; produced by a gland

Hydrocarbon – a compound containing only hydrogen and carbon atoms

Infectious – a disease that can be transferred from one person to another

Insoluble – a substance that will not dissolve in a solvent

Insulin – a hormone that controls blood glucose concentration

Ion – a particle that has a positive or negative electrical charge

Ion (mineral) – a mineral found as a charged particle, e.g. Ca^{2+}, that the body needs to function properly

Ionic bond – formed when two or more atoms gain or lose electrons to become charged ions

Ionising power – the ability to ionise other atoms

Isotopes – atoms of the same element but with a different number of neutrons

Leprosy – a contagious bacterial disease affecting the skin and nerves

Longitudinal wave – causes particles in the medium to move backwards and forwards in the direction in which the wave is moving

Mass number – the total number of protons and neutrons present in an atom

Methane – a colourless, odourless inflammable gas; a greenhouse gas

Microorganism – a very small organism

Mineral – a naturally occurring chemical element or compound found in rocks

Mitochondria – where cell respiration takes place

MRSA – an antibiotic-resistant bacterium; 'superbug'

Mutation – a change in the genetic material of a cell

Nanoscience – dealing with materials that have a very small grain size, in the order of 1–100nm

National Grid – network of cables that transfers electricity around the country

Natural selection – the survival of individual organisms that are best suited / adapted to their environment

Neurone – a specialised cell that transmits electrical impulses

Neutron – a subatomic particle found in the nucleus; has no charge

Non-biodegradable – a substance that does not decompose naturally by the action of microorganisms

Non-renewable – energy sources that cannot be replaced in a lifetime

Nuclear – energy obtained from the nucleus of an atom

Nucleus – controls the activity of the cell; centre of an atom containing protons and neutrons

Ore – a naturally occurring mineral, from which it is economically viable to extract a metal

Pathogen – a disease-causing microorganism

Pesticide – a substance used for destroying insects and other pests

Photosynthesis – the chemical process that uses light energy to produce glucose in plants

Phototropism – a plant's growth response to light

Phytomining – using plants to remove metals from soil and the ground

Pituitary gland – the small gland at the base of the brain that produces hormones

Pollution – the contamination of an environment by chemicals, waste or heat

Polymer – a giant, long-chained molecule

Polymerisation – the process of monomers joining together to form a polymer

Power – the rate of doing work; measured in watts

Precipitation – the removal of particles from a solution

Product – the substance made at the end of a chemical reaction

Proton – a positively charged subatomic particle in the nucleus

Radiation – electromagnetic particles / rays emitted by a radioactive substance

Reactant – a substance present before a chemical reaction takes place

Receptor – a sense organ, e.g. eyes, ears, nose, etc

Red-shift – the shift of light towards the red part of the visible spectrum; suggesting that the Universe is expanding

Reduction – a reaction involving the loss of oxygen, the addition of hydrogen or the gain of electrons

Reflection – a wave (e.g. light or sound) that is thrown back from a surface

Reflex action – an involuntary action, e.g. automatically jerking your hand away from something hot

Refraction – the change in direction of a wave as it passes from one medium to another

Relative atomic mass, A$_r$ – the average mass of an atom of an element compared with one twelfth of a carbon atom

Relative formula mass, M$_r$ – the sum of the atomic masses of all atoms in a molecule

Renewable energy – energy sources that can be replaced

Resistor – a electrical device that resists the flow of an electrical current

Respiration – the process of releasing energy from glucose in cells

Ribosome – part of a cell that makes proteins from amino acids

Salt – the product of a chemical reaction between a base and an acid

Sedimentary rock – rock formed by the accumulation of sediment

Side effect – condition caused by taking medication, e.g. headache, nausea

Smart material – a material that has one or more properties that can be altered

Specialised – developed for a special function

Stem cell – a cell of human embryos or adult bone marrow that has yet to differentiate

Sustainable – resources that can be replaced

Synapse – the gap between two neurones

Tectonic plate – huge sections of the Earth's crust that move relative to one another

Telescope – a device that magnifies distant objects

Theory – the best way to explain why something is happening. It can be changed when new evidence is found

Thermal decomposition – when heat is used to break down a chemical compound

Thermal energy – heat energy

Thermoregulation – maintenance of a constant body temperature in warm-blooded animals

Toxin – a poison

Transfer – to move energy from one place to another

Transform – to change energy from one form into another, e.g. electrical energy to heat energy

Transformer – an electrical device used to change the voltage of alternating currents

Transmission – the sending of information / electricity / electromagnetic radiation over a communications line or circuit

Tsunami – a huge wave caused by an earthquake under the sea

Vaccine – a liquid preparation used to make the body produce antibodies to provide protection against disease

Variation – difference between individuals of the same species

Voltage – potential difference; the difference in electrical charge between two charged points expressed in volts

X-ray – a type of electromagnetic radiation

Yield – the amount of product obtained from a reaction

Zygote – a cell formed by the fusion of the nuclei of a male sex cell and a female sex cell (gametes)

Index

Acids 46
Adaptation 35
Air 26
Alcohol 72
Alkalis 46, 47
Alkanes 25, 58
Alkenes 56
Alleles 50
Alloy 27
Alpha particles 74
Anode 78
Antibiotics 71
Atom economy 32
Atomic number 20
Atoms 20

Background radiation 75
Bases 47
Beta particles 74
Big Bang 13
Biodiversity 84
Biomass, pyramids of 37
Brain 42
Breeding, selective 81

Cancer 72, 75, 76
Carbon cycle 39
Carbon dioxide 17, 84
Cathode 78
Cell membrane 48
Cells 48
Cement 53
Central nervous system 42, 43
Ceramics 54, 55
Chemical formulae 22
Chemical reactions 30
Chromic materials 80
Chromosomes 48, 49, 52
Classification 34
Clones 81
Coal 60
Community 34
Competition 35

Composite materials 54, 55
Compounds 22
Concrete 53
Conduction 87
Continental drift 14
Convection 87
Core 14
Covalent bonds 54
Crude oil 25, 60
Crust 14
Cytoplasm 48

Decay 38
Deforestation 84
Development, sustainable 85
Diabetes 44
Differentiation 51
Distillation 25
DNA 52
Dominant alleles 50
Doppler effect 13
Drugs 70, 72

Earth's atmosphere 17–18
Earth's structure 14
Earthquakes 16
Effectors 42
Efficiency 65
Electrical power 64
Electrolysis 77
Electromagnetic spectrum 66
Electron configuration 21
Electrons 20
Electroplating 78
Elements 20
Empirical formulae 29
Energy transfer 38, 65
Energy transformation 64
Eutrophication 85
Evolution, theory of 35, 36
Extinction 36
Extraction methods 27

Feedback, negative 45
Food chains 37
Fossil fuels 18, 58, 59
Fossils 14, 35
Fractional distillation 25
Fuels 58, 59

Gametes 49
Gamma rays 66–67, 74
Genes 48, 52, 81
Genetic disorders 52
Genetic engineering 81, 82
Glands 44
Global warming 19, 58, 86
Glucagon 44
Gravitropism 36
Greenhouse effect 19, 58, 84

Hazard symbols 46
Heterozygous 50
Homeostasis 44
Homozygous 50
Hormones 44
Hydrocarbons 25

Immunity 73
Infrared rays 66–67, 87
Insulin 44, 82
Ionic bonds 54
Ions 22, 77
Iron 27, 28
Isotopes 23, 74

Kingdoms 34

Lead 28
Limestone 39, 53

Mantle 14
Mass number 20
Materials 24
Metals 54
Microorganisms 38, 71
Microwaves 66–67

Minerals 26
Mitochondria 48
Mixtures 22, 24
Monohybrid inheritance 51
Mortar 53
Mutation 36

Nanoscience 79
National Grid 63
Natural gas 60
Natural selection 36, 71
Neutrons 20
Nitrogen 17
Non-renewable energy 83
Nuclear fuels 59, 60
Nucleus 48

Ores 26
Oxygen 17

Pathogens 71, 73
Periodic table 21
Photosynthesis 39
Phototropism 36
Phytomining 26
Plastics 56
Platelets 73
Pollution 83, 89
Polymerisation 56–57
Polymers 57
Population 34
Precipitation 47
Products 30–31
Protons 20

Radiation 74, 76, 87
Radio waves 66–67
Reactants 30–31
Receptors 42, 43
Recessive alleles 50
Red blood cells 73
Red-shift 13
Reflecting telescope 12
Reflection 66

Reflex actions 43
Refracting telescope 12
Refraction 66
Relative atomic number 23
Relative mass number 23
Renewable energy 61–62
Respiration 39, 48
Ribosomes 48
Rock cycle 14

Salts 47
Sankey diagrams 65
Sedimentary rocks 18
Smart materials 57, 79–80
Species 34
Spinal cord 42
Spinal nerves 42
State symbols 77
Stem cells 52
Superconductors 80
Synapse 42

Tectonic plates 15–16
Tectonic theory 14
Telescopes 12
Thermal decomposition 53
Thermoregulation 45
Tobacco 72
Tsunamis 16

Ultraviolet rays 66–67

Vaccination 73
Variation 36, 48
Visible light 66–67
Volcanoes 16

Wegener, Alfred 14
White blood cells 71, 73

X-ray telescope 12
X-rays 66–67

Yield 32, 33

Zygotes 49

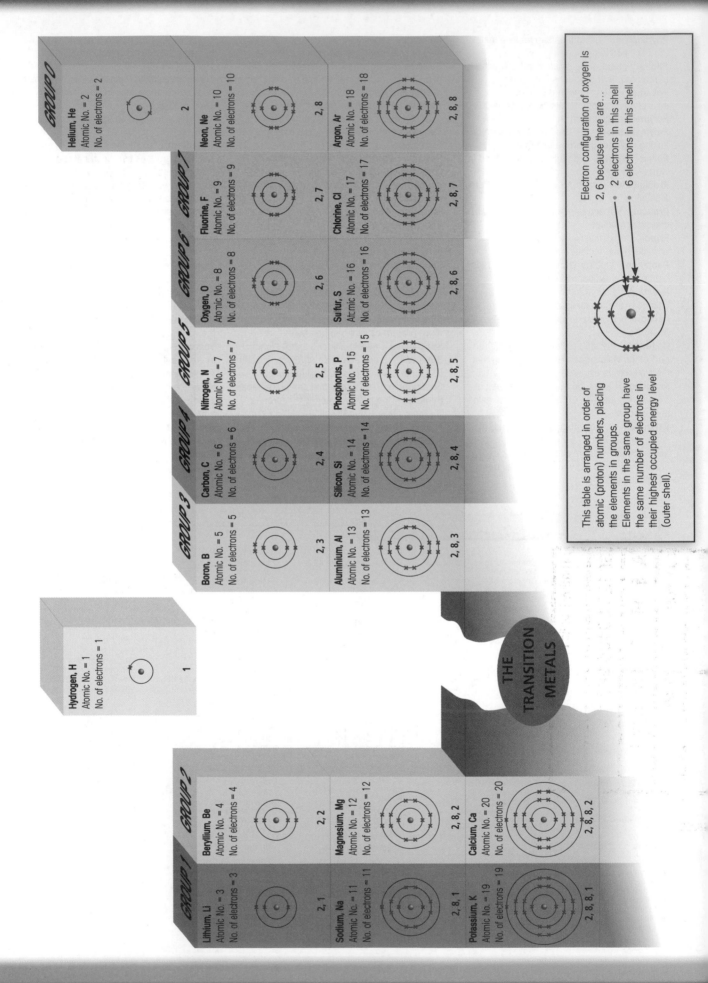

GROUP 0

Helium, He
Atomic No. = 2
No. of electrons = 2

2

Neon, Ne
Atomic No. = 10
No. of electrons = 10

2, 8

Argon, Ar
Atomic No. = 18
No. of electrons = 18

2, 8, 8

GROUP 7

Fluorine, F
Atomic No. = 9
No. of electrons = 9

2, 7

Chlorine, Cl
Atomic No. = 17
No. of electrons = 17

2, 8, 7

GROUP 6

Oxygen, O
Atomic No. = 8
No. of electrons = 8

2, 6

Sulfur, S
Atomic No. = 16
No. of electrons = 16

2, 8, 6

GROUP 5

Nitrogen, N
Atomic No. = 7
No. of electrons = 7

2, 5

Phosphorus, P
Atomic No. = 15
No. of electrons = 15

2, 8, 5

GROUP 4

Carbon, C
Atomic No. = 6
No. of electrons = 6

2, 4

Silicon, Si
Atomic No. = 14
No. of electrons = 14

2, 8, 4

GROUP 3

Boron, B
Atomic No. = 5
No. of electrons = 5

2, 3

Aluminium, Al
Atomic No. = 13
No. of electrons = 13

2, 8, 3

Hydrogen, H
Atomic No. = 1
No. of electrons = 1

1

THE TRANSITION METALS

GROUP 1

Lithium, Li
Atomic No. = 3
No. of electrons = 3

2, 1

Sodium, Na
Atomic No. = 11
No. of electrons = 11

2, 8, 1

Potassium, K
Atomic No. = 19
No. of electrons = 19

2, 8, 8, 1

GROUP 2

Beryllium, Be
Atomic No. = 4
No. of electrons = 4

2, 2

Magnesium, Mg
Atomic No. = 12
No. of electrons = 12

2, 8, 2

Calcium, Ca
Atomic No. = 20
No. of electrons = 20

2, 8, 8, 2

Electron configuration of oxygen is 2, 6 because there are...
• 2 electrons in this shell
• 6 electrons in this shell.

This table is arranged in order of atomic (proton) numbers, placing the elements in groups.
Elements in the same group have the same number of electrons in their highest occupied energy level (outer shell).